Elements of
Abstract Algebra

Allan Clark
Purdue University

Dover Publications, Inc.
New York

For my parents

I was just going to say, when I was interrupted, that one of the many ways of classifying minds is under the heads of arithmetical and algebraical intellects. All economical and practical wisdom is an extension of the following arithmetical formula: $2 + 2 = 4$. Every philosophical proposition has the more general character of the expression $a + b = c$. We are mere operatives, empirics, and egotists until we learn to think in letters instead of figures.

OLIVER WENDELL HOLMES
The Autocrat of the Breakfast Table

Published in Canada by General Publishing Company, Ltd., 30 Lesmill Road, Don Mills, Toronto, Ontario.

Published in the United Kingdom by Constable and Company, Ltd., 10 Orange Street, London WC2H 7EG.

This Dover edition, first published in 1984, is an unabridged and corrected republication of the work first published by Wadsworth Publishing Company, Belmont, California, in 1971.

Manufactured in the United States of America
Dover Publications, Inc., 31 East 2nd Street, Mineola, N.Y. 11501

Library of Congress Cataloging in Publication Data

Clark, Allan, 1935–
   Elements of abstract algebra.

   "Corrected republication"—Verso t.p.
   Originally published: Belmont, Calif. : Wadsworth, © 1971.
   Bibliography: p.
   Includes index.
   1. Algebra, Abstract.   I. Title.
[QA162.C57   1984]   512'.02   84-6118
ISBN 0-486-64725-0

Foreword

Modern or "abstract" algebra is widely recognized as an essential element of mathematical education. Moreover, it is generally agreed that the axiomatic method provides the most elegant and efficient technique for its study. One must continually bear in mind, however, that the axiomatic method is an organizing principle and not the substance of the subject. A survey of algebraic structures is liable to promote the misconception that mathematics is the study of axiom systems of arbitrary design. It seems to me far more interesting and profitable in an introductory study of modern algebra to carry a few topics to a significant depth. Furthermore I believe that the selection of topics should be firmly based on the historical development of the subject.

This book deals with only three areas of abstract algebra: group theory, Galois theory, and classical ideal theory. In each case there is more depth and detail than is customary for a work of this type. Groups were the first algebraic structure characterized axiomatically. Furthermore the theory of groups is connected historically and mathematically to the Galois theory of equations, which is one of the roots of modern algebra. Galois theory itself gives complete answers to classical problems of geometric constructibility and solvability of equations in radicals. Classical ideal theory, which arose from the problems of unique factorization posed by Fermat's last theorem, is a natural sequel to Galois theory and gives substance to the study of rings. All three topics converge in the fundamental theorem of algebraic number theory for Galois extensions of the rational field, the final result of the book.

Emil Artin wrote: *We all believe that mathematics is an art. The author of a book, the lecturer in a classroom tries to convey the structural beauty of mathematics to his readers, to his listeners. In this attempt he must always fail. Mathe-*

Three   Field Theory

Four   Galois Theory

Five   Ring Theory

Six   Classical Ideal Theory

Introduction

Classical algebra was the art of resolving equations. Modern algebra, the subject of this book, appears to be a different science entirely, hardly concerned with equations at all. Yet the study of abstract structure which characterizes modern algebra developed quite naturally out of the systematic investigation of equations of higher degree. What is more, the modern abstraction is needed to bring the classical theory of equations to a final perfect form.

The main part of this text presents the elements of abstract algebra in a concise, systematic, and deductive framework. Here we shall trace in a leisurely, historical, and heuristic fashion the genesis of modern algebra from its classical origins.

The word *algebra* comes from an Arabic word meaning "reduction" or "restoration." It first appeared in the title of a book by Muhammad ibn Musa al-Khwarizmi about the year 825 A.D. The renown of this work, which gave complete rules for solving quadratic equations, led to use of the word algebra for the whole science of equations. Even the author's name lives on in the word *algorithm* (a rule for reckoning) derived from it. Up to this point the theory of equations had been a collection of isolated cases and special methods. The work of al-Khwarizmi was the first attempt to give it form and unity.

The next major advance came in 1545 with the publication of *Artis Magnae sive de Regulis Algebraicis* by Hieronymo Cardano (1501–1576). Cardano's book, usually called the *Ars Magna*, or "The Grand Art," gave the complete solution of equations of the third and fourth degree. Exactly how much credit for these discoveries is due to Cardano himself we cannot be certain. The solution of the quartic is due to Ludovico Ferrari (1522–1565), Cardano's student, and the solution of the cubic was based in part upon earlier work of Scipione del Ferro (1465?–1526). The claim of Niccolo Fontana (1500?–1557), better known as Tartaglia ("the stammerer"), that he gave Cardano the cubic under a pledge of secrecy, further complicates the issue. The bitter feud between Cardano and Tartaglia obscured the true primacy of del Ferro.

A solution of the cubic equation leading to Cardano's formula is quite simple to give and motivates what follows. The method we shall use is due to Hudde, about 1650. Before we start, however, it is necessary to recall that *every complex number has precisely three cube roots*. For example, the complex number $1 = 1 + 0i$ has the three cube roots, 1 (itself), $\omega = -\frac{1}{2} + \frac{1}{2}\sqrt{-3}$, and $\omega^2 = -\frac{1}{2} - \frac{1}{2}\sqrt{-3}$. In general, if z is any one of the cube roots of a complex number w, then the other two are ωz and $\omega^2 z$.

For simplicity we shall consider only a special form of the cubic equation,

$$x^3 + qx - r = 0. \tag{1}$$

(However, the general cubic equation may always be reduced to one of this form without difficulty.) First we substitute $u + v$ for x to obtain a new equation,

$$(u^3 + 3u^2v + 3uv^2 + v^3) + q(u + v) - r = 0, \tag{2}$$

which we rewrite as

$$u^3 + v^3 + (3uv + q)(u + v) - r = 0. \tag{3}$$

Since we have substituted two variables, u and v, in place of the one variable x, we are free to require that $3uv + q = 0$, or in other words, that $v = -q/3u$. We use this to eliminate v from (3), and after simplification we obtain,

$$u^6 - ru^3 - \frac{q^3}{27} = 0. \tag{4}$$

This last equation is called the *resolvent equation* of the cubic (1). We may view it as a quadratic equation in u^3 and solve it by the usual method to obtain

$$u^3 = \frac{r}{2} \pm \sqrt{\frac{r^2}{4} + \frac{q^3}{27}}. \tag{5}$$

Of course a complete solution of the two equations embodied in (5) gives six values of u—three cube roots for each choice of sign. These six values of u are the roots of the sixth-degree resolvent (4). We observe however that if u is a cube root of $(r/2) + \sqrt{(r^2/4) + (q^3/27)}$, then $v = -q/3u$ is a cube root of $(r/2) - \sqrt{(r^2/4) + (q^3/27)}$. Consequently the six roots of (4) may be conveniently designated as u, ωu, $\omega^2 u$ and v, ωv, $\omega^2 v$, where $uv = -q/3$. Thus the three roots of the original equation are

$$\alpha_1 = u + v, \qquad \alpha_2 = \omega u + \omega^2 v, \qquad \alpha_3 = \omega^2 u + \omega v, \tag{6}$$

where

$$u^3 = \frac{r}{2} + \sqrt{\frac{r^2}{4} + \frac{q^3}{27}} \quad \text{and} \quad v = \frac{-q}{3u}.$$

In other words, the roots of the original cubic equation (1) are given by the *formula of Cardano*,

$$\alpha = \sqrt[3]{\frac{r}{2} + \sqrt{\frac{r^2}{4} + \frac{q^3}{27}}} + \sqrt[3]{\frac{r}{2} - \sqrt{\frac{r^2}{4} + \frac{q^3}{27}}},$$

in which the cube roots are varied so that their product is always $-q/3$.

For our purposes we do not need to understand fully this complete solution of the cubic equation—only the general pattern is of interest here. The important fact is that the roots of the cubic equation can be expressed in terms of the roots of a resolvent equation which we know how to solve. The same fact is true of the general equation of the fourth degree.

For a long time mathematicians tried to find a solution of the general quintic, or fifth-degree, equation without success. No method was found to carry them beyond the writings of Cardano on the cubic and quartic. Consequently they turned their attention to other aspects of the theory of equations, proving theorems about the distribution of roots and finding methods of approximating roots. In short, the theory of equations became analytic.

One result of this approach was the discovery of the *fundamental theorem of algebra* by D'Alembert in 1746. The fundamental theorem states that every algebraic equation of degree n has n roots. It implies, for example, that the equation $x^n - 1 = 0$ has n roots—the so-called *nth roots of unity*—from which it follows that every complex number has precisely n nth roots. D'Alembert's proof of the fundamental theorem was incorrect (Gauss gave the first correct proof in 1799) but this was not recognized for many years, during which the theorem was popularly known as " D'Alembert's theorem."

D'Alembert's discovery made it clear that the question confronting algebraists was not the existence of solutions of the general quintic equation, but whether or not the roots of such an equation could be expressed in terms of its coefficients by means of formulas like those of Cardano, involving only the extraction of roots and the rational operations of addition, subtraction, multiplication, and division.

In a new attempt to resolve this question Joseph Louis Lagrange (1736–1813) undertook a complete restudy of all the known methods of solving cubic and quartic equations, the results of which he published in 1770 under the title *Réflexions sur la résolution algébrique des equations*. Lagrange observed that the roots of the resolvent equation of the cubic (4) can be expressed in terms of the roots α_1, α_2, α_3 of the original equation (1) in a completely symmetric fashion. Specifically,

$$v = \tfrac{1}{3}(\alpha_1 + \omega\alpha_2 + \omega^2\alpha_3), \qquad u = \tfrac{1}{3}(\alpha_1 + \omega\alpha_3 + \omega^2\alpha_2),$$
$$\omega v = \tfrac{1}{3}(\alpha_3 + \omega\alpha_1 + \omega^2\alpha_2), \qquad \omega u = \tfrac{1}{3}(\alpha_2 + \omega\alpha_1 + \omega^2\alpha_3), \qquad (7)$$
$$\omega^2 v = \tfrac{1}{3}(\alpha_2 + \omega\alpha_3 + \omega^2\alpha_1), \qquad \omega^2 u = \tfrac{1}{3}(\alpha_3 + \omega\alpha_2 + \omega^2\alpha_1).$$

All these expressions may be obtained from any one of them by permuting the occurrences of α_1, α_2, α_3 in all six possible ways.

Lagrange's observation was important for several reasons. We obtained the resolvent of the cubic by making the substitution $x = u + v$. Although this works quite nicely, there is no particular rhyme nor reason to it—it is definitely *ad hoc*. However Lagrange's observation shows how we might have constructed the resolvent on general principles and suggests a method for constructing resolvents of equations of higher degrees. Furthermore it shows that the original equation is solvable in radicals if and only if the resolvent equation is.

To be explicit let us consider a quartic equation,

$$x^4 - px^3 + qx^2 - rx + s = 0, \qquad (8)$$

and suppose that the roots are the unknown complex numbers α_1, α_2, α_3, α_4. Without giving all the details we shall indicate how to construct the resolvent equation. First we recall that the fourth roots of unity are the complex numbers 1, i, i^2, i^3, where $i = \sqrt{-1}$ and $i^2 = -1$, $i^3 = -i$. Then the roots of the resolvent are the twenty-four complex numbers

$$u_{ijkl} = \tfrac{1}{4}(\alpha_i + i\alpha_j + i^2\alpha_k + i^3\alpha_l), \qquad (9)$$

where the indices i, j, k, l are the numbers 1, 2, 3, 4 arranged in some order. Therefore the resolvent equation is the product of the twenty-four distinct factors $(x - u_{ijkl})$. That is, we may write the resolvent equation in the form

$$\phi(x) = \prod_{ijkl} (x - u_{ijkl}) = 0. \qquad (10)$$

Thus the resolvent of the quartic has degree 24, and it would seem hopeless to solve. It turns out, however, that every exponent of x in $\phi(x)$ is divisible by

4, and consequently $\phi(x) = 0$ may be viewed as a sixth-degree equation in x^4. What is more, this sixth-degree equation can be reduced to the product of two cubic equations (in a way we cannot make explicit here). Since cubics can be solved, a solution of the quartic can be obtained by a specific formula in radicals. (Such a formula is so unwieldy that it is more useful and understandable simply to describe the process for obtaining solutions.)

For quintic, or fifth-degree, equations Lagrange's theory yields a resolvent equation of degree 120, which is a 24th-degree equation in x^5. Lagrange was convinced that his approach, which revealed the similarities in the resolution of cubics and quartics, represented the true metaphysics of the theory of equations. The difficulty of the computations prevented Lagrange from testing whether his techniques could produce a formula for resolving the quintic in radicals. Moreover, with his new insights, Lagrange could foresee the point at which the process might break down, and he gave equal weight to the impossibility of such a formula.

A short time afterward, Paolo Ruffini (1765–1822) published a proof of the unsolvability of quintic equations in radicals. Ruffini's argument, given in his two-volume *Teoria generale delle equazioni* of 1799, was correct in essence, but was not, in actual fact, a proof. A complete and correct proof was given by Niels Henrik Abel (1802–1829) in 1826 in a small book published at his own expense. The brilliant work of Abel closed the door on a problem which had excited and frustrated the best mathematical minds for almost three centuries.

There remained one final step. Some equations of higher degree are clearly solvable in radicals even though they cannot be factored. Abel's theorem raised the question: which equations are solvable in radicals and which are not? The genius Évariste Galois (1811–1832) gave a complete answer to this question in 1832. Galois associated to each algebraic equation a system of permutations of its roots, which he called a *group*. He was able to show equivalence of the solvability of an equation in radicals, with a property of its group. Thus he made important discoveries in the theory of groups as well as the theory of equations. Unfortunately Galois' brief and tragic life ended in a foolish duel before his work was understood. His theory perfected the ideas of Lagrange, Ruffini, and Abel and remains one of the stunning achievements of modern mathematical thought.

At this point we can only leave as a mystery the beautiful relation Galois discovered between the theory of equations and the theory of groups—a mystery resolved by the deep study of both theories undertaken in the text.

We can, however, gain some insight into modern abstraction by a short and informal discussion of groups. To take an example near at hand, we shall consider the group of permutations of the roots α_1, α_2, α_3 of the cubic equation—which happens to be the Galois group of this equation in general. This group consists of six operations, A, B, C, D, E, and I, specified as follows:

A leaves α_1 fixed and interchanges the roots α_2 and α_3 wherever they occur.
B leaves α_2 fixed and interchanges α_1 and α_3.
C interchanges α_1 and α_2, leaving α_3 fixed.
D replaces α_1 by α_2 at each occurrence, α_2 by α_3, and α_3 by α_1.
E replaces α_1 by α_3, α_3 by α_2, and α_2 by α_1.
I is the identity operation, which makes no change at all.

For example, the result of applying the operation A to v, as expressed in (7), is u. We indicate this by writing

$$A(v) = u.$$

Similarly, the result of applying the operation E to v is ωv, or in other words, $E(v) = \omega v$. Of course, by definition, $I(v) = v$. It is easy to verify that by applying the six operations A, B, C, D, E, and I to v, we obtain all six of the expressions in (7) for the roots of the resolvent equation.

These operations have the property that if any two of them are applied successively, the result is the same as if one of the others had been applied once. For example, suppose we apply the operation A to v, obtaining u, and then apply the operation D to u, obtaining ωu. The result is the same as if we had applied the operation C directly to v. We can express this in symbols by

$$D(A(v)) = C(v).$$

In fact this remains true no matter what we put in place of v. That is, the result of first applying the operation A and then applying D is the same as applying the operation C. We sum this up in the simple equation: $DA = C$. There are many other relations of this sort among these operations. For example, we may compute the result of the composite operation EB on any function $f(\alpha_1, \alpha_2, \alpha_3)$ as follows:

$$B(f(\alpha_1, \alpha_2, \alpha_3)) = f(\alpha_3, \alpha_2, \alpha_1),$$
$$EB(f(\alpha_1, \alpha_2, \alpha_3)) = E(f(\alpha_3, \alpha_2, \alpha_1)) = f(\alpha_2, \alpha_1, \alpha_3) = C(f(\alpha_1, \alpha_2, \alpha_3)).$$

Thus $EB = C$. The thirty-six relations of this type can be given conveniently in a table. We put the result of the composite operation XY in the X row and the Y column.

We observe now that composition of the operations A, B, C, D, E, and I has the following properties.

(1) For any three operations X, Y, and Z, we have

$$X(YZ) = (XY)Z.$$

In other words, the result of first performing the operation YZ and then the operation X is the same as the result of first performing the operation Z and

Table 1

	A	B	C	D	E	I
A	I	D	E	B	C	A
B	E	I	D	C	A	B
C	D	E	I	A	B	C
D	C	A	B	E	I	D
E	B	C	A	I	D	E
I	A	B	C	D	E	I

then the operation XY. For example, from Table 1 we see that $AB = D$ and $BC = D$, and therefore

$$A(BC) = AD = B = DC = (AB)C.$$

Thus we have verified the equation above for the special case where $X = A$, $Y = B$, and $Z = C$. This property of the composition of the operations is called *associativity*. To verify associativity completely from Table 1 we would have to make 216 checks like the one above.

(2) For any operation X we have

$$XI = X = IX.$$

In other words, the composition of any operation X with the identity operation I always gives X again. This property is easily checked by examining the last row and the last column of Table 1.

(3) For any operation X there is precisely one operation Y such that

$$XY = I = YX.$$

In other words, whatever the operation X does to the roots $\alpha_1, \alpha_2, \alpha_3$, Y does just the opposite. We call Y the inverse of X and denote it by X^{-1}. It is easy to see from Table 1 that

$$A^{-1} = A, \quad B^{-1} = B, \quad C^{-1} = C, \quad D^{-1} = E, \quad E^{-1} = D, \quad I^{-1} = I.$$

Whenever we have a set of operations and a rule for composing them that satisfies these three properties, we say that the operations form a *group*.

Once we know that a set of operations with a particular rule for composing them is a group, we can analyze properties of these operations and their composition without regard to the manner in which they are defined or the context in which they arose. This simplifies the situation by eliminating irrelevant details, and gives the work generality.

To clarify this process of abstraction, let us consider another group of operations defined in a completely different way. Again we shall have six operations, but this time we shall call them by the Greek letters α, β, γ, δ, ε, and ι. These will operate on the rational numbers (except 0 and 1) by the following rules:

$$\alpha(x) = \frac{1}{x}, \qquad \delta(x) = \frac{1}{1-x},$$

$$\beta(x) = 1 - x, \qquad \varepsilon(x) = \frac{x-1}{x},$$

$$\gamma(x) = \frac{x}{x-1}, \qquad \iota(x) = x,$$

where x is any rational number except 0 or 1. We may compose these operations and the result will always be one of the other operations. For example, we have that $\delta\alpha = \gamma$, since

$$\delta(\alpha(x)) = \delta\left(\frac{1}{x}\right) = \frac{1}{1-(1/x)} = \frac{x}{x-1} = \gamma(x).$$

Again, we may make a table of all thirty-six compositions of these six operations.

Table 2

	α	β	γ	δ	ε	ι
α	ι	δ	ε	β	γ	α
β	ε	ι	δ	γ	α	β
γ	δ	ε	ι	α	β	γ
δ	γ	α	β	ε	ι	δ
ε	β	γ	α	ι	δ	ε
ι	α	β	γ	δ	ε	ι

It is immediately apparent that Table 2 has a strong resemblance to Table 1. For example, every occurrence of A in the first table corresponds to an occurrence of α in the second. Similarly the letters B and β occur in the same positions in each table. In fact Table 1 may be transformed into Table 2 by making the substitutions:

$$A \to \alpha, \quad B \to \beta, \quad C \to \gamma, \quad D \to \delta, \quad E \to \varepsilon, \quad I \to \iota.$$

In other words, these two groups have the same structure *as groups* even though the individual operations are defined in quite different ways. To put it another way, all the facts which depend solely upon the way operations are composed will be the same for both groups. In such a case two groups are said to be *isomorphic*. Group theory studies the properties of groups which remain unchanged in passing from one group to another isomorphic with it.

Group theory was called the "theory of substitutions" until 1854 when the English mathematician Arthur Cayley (1821–1895) introduced the concept of *abstract group*. The convenience and power of the abstract approach to group theory was evident by the end of the nineteenth century. Subsequent abstractions, such as *field* and *ring*, have also proved to be powerful concepts. The success of abstract thinking in algebra has been so enormous that the terms *modern algebra* and *abstract algebra* are synonymous.

Abstraction is simply the process of separating form from content. We abstract whenever we pass from a particular instance to the general case. Even the simplest mathematics, ordinary arithmetic, is an abstraction from physical reality. In modern mathematics we abstract from previous mathematical experience and reach a new and higher plane of abstraction. Indeed, each mathematical generation abstracts from the work of preceding ones, continually distilling and concentrating the essence of old thought into new and more perfect forms. The rewards are great. Not only does abstraction greatly enhance our understanding, it also dramatically increases the applications of mathematics to practical life. Even such an apparently recondite subject as group theory has applications in crystallography and quantum mechanics. Over centuries modern algebra has grown into a large body of abstract knowledge worthy of study both for its intrinsic fascination and extrinsic application.

Chapter 1

Set theory is the proper framework for abstract mathematical thinking. All of the abstract entities we study in this book can be viewed as sets with specified additional structure. Set theory itself may be developed axiomatically, but the goal of this chapter is simply to provide sufficient familiarity with the notation and terminology of set theory to enable us to state definitions and theorems of abstract algebra in set-theoretic language. It is convenient to add some properties of the natural numbers to this informal study of set theory.

It is well known that an informal point of view in the theory of sets leads to contradictions. These difficulties all arise in operations with very large sets. We shall never need to deal with any sets large enough to cause trouble in this way, and, consequently, we may put aside all such worries.

The Notation and
Terminology of Set Theory

---

1.   A *set* is any aggregation of objects, called *elements* of the set. Usually the elements of a set are mathematical quantities of a uniform character. For example, we shall have frequent occasion to consider the set of integers $\{\ldots, -2, -1, 0, 1, 2, \ldots\}$, which is customarily denoted **Z** (for the German "Zahlen," which means "numbers"). We shall use also the set **Q** of rational numbers—numbers which are the quotient of two integers, such as $7/3$, $-4/5$, 2.

To give an example of another type, we let K denote the set of coordinate points (x, y) in the xy-coordinate plane such that $x^2 + y^2 = 1$. Then K is the circle of unit radius with the origin as center.

2.   To indicate that a particular quantity x is an element of the set S, we write $x \in S$, and to indicate that it is not, we write $x \notin S$. Thus $-2 \in \mathbf{Z}$, but $1/2 \notin \mathbf{Z}$; and $1/2 \in \mathbf{Q}$, but $\sqrt{2} \notin \mathbf{Q}$.

A set is completely determined by its elements. Two sets are equal if and only if they have precisely the same elements. In other words, $S = T$ if and only if $x \in S$ implies $x \in T$ and $x \in T$ implies $x \in S$.

It will be convenient to write $x, y, z \in S$ for $x \in S$, $y \in S$, and $z \in S$.

3.   A set S is a *subset* of a set T if every element of S is an element of T, or in other words, if $x \in S$ implies $x \in T$. To indicate that S is a subset of T we write $S \subset T$. If $S \subset T$ and $T \subset S$, then $x \in S$ implies $x \in T$ and $x \in T$ implies $x \in S$, so that $S = T$.

The *empty set* $\emptyset$ is the set with no elements whatever. The empty set is a subset of every set T. If S is a subset of T and neither $S = \emptyset$ nor $S = T$, then S is called a *proper subset* of T.

4.   Frequently a set is formed by taking for its elements all objects which have a specific property. We shall denote the set of all x with the property P by $\{x \mid P(x)\}$. Thus,

$$\mathbf{Z} = \{x \mid x \text{ is an integer}\}.$$

To indicate that a set is formed by selecting from a given set S those elements with property P, we write $\{x \in S \mid P(x)\}$. It is clear that $\{x \in S \mid P(x)\}$ is always a subset of S. For example, the set of even integers,

$$2\mathbf{Z} = \{x \in \mathbf{Z} \mid x = 2y, y \in \mathbf{Z}\},$$

is a subset of **Z**.

5. The *intersection* of two sets S and T is the set $S \cap T$ of elements common to both. In other words,

$$S \cap T = \{x \mid x \in S \text{ and } x \in T\}.$$

The intersection $S \cap T$ is a subset of both S and T. The sets S and T are said to be *disjoint* if $S \cap T = \emptyset$.

We note the following properties of intersection:

(a) $A \cap (B \cap C) = (A \cap B) \cap C$,
(b) $A \cap B = B \cap A$,
(c) $A \cap A = A$ and $A \cap \emptyset = \emptyset$,
(d) $A \cap B = A$ if and only if $A \subset B$.

Let $S_1, S_2, \ldots, S_n$ be sets. Then we shall write

$$\bigcap_{i=1}^{n} S_i$$

as an abbreviation for

$$S_1 \cap S_2 \cap \cdots \cap S_n = \{x \mid x \in S_i \text{ for each } i = 1, 2, \ldots, n\}.$$

6. The *union* of two sets S and T is the set $S \cup T$ of elements in S or T or in both S and T. In other words,

$$S \cup T = \{x \mid x \in S \text{ and/or } x \in T\}.$$

S and T are both subsets of $S \cup T$.

The following properties of union are analogous to those of intersection:

(a) $A \cup (B \cup C) = (A \cup B) \cup C$,
(b) $A \cup B = B \cup A$,
(c) $A \cup A = A$ and $A \cup \emptyset = A$,
(d) $A \cup B = B$ if and only if $A \subset B$.

Let $S_1, S_2, \ldots, S_n$ be sets. Then we shall write $\bigcup_{i=1}^{n} S_i$ as an abbreviation for

$$S_1 \cup S_2 \cup \cdots \cup S_n = \{x \mid x \in S_i \text{ for at least one } i = 1, 2, \ldots, n\}.$$

7. Intersection and union are related by the following *distributive laws*:

(a) $A \cup (B \cap C) = (A \cup B) \cap (A \cup C)$,
(b) $A \cap (B \cup C) = (A \cap B) \cup (A \cap C)$.

8. The *difference* of two sets S and T is the set $S - T$ of elements of S which are not elements of T. In other words,

$$S - T = \{x \in S \mid x \notin T\}.$$

$S - T$ is always a subset of S.

The difference of sets has the following properties:

(a) $A - B = \emptyset$ if and only if $A \subset B$,
(b) $A - B = A$ if and only if $A \cap B = \emptyset$,
(c) $A - B = A - C$ if and only if $A \cap B = A \cap C$,
(d) $A - \emptyset = A$ and $A - A = \emptyset$,
(e) $A - (B \cap C) = (A - B) \cup (A - C)$,
(f) $A - (B \cup C) = (A - B) \cap (A - C)$.

8α.   The *symmetric difference* of two sets A and B is the set

$$A * B = (A - B) \cup (B - A).$$

Show that $A * B = (A \cup B) - (A \cap B)$. Show that $A * B = \emptyset$ if and only if $A = B$. Prove that the symmetric difference is an associative operation on sets, that is to say, $A * (B * C) = (A * B) * C$ for any three sets A, B, and C.

8β.   If every set in a discussion is a subset of a given set $\mathfrak{A}$, then we call $\mathfrak{A}$ the *universe* (of that discussion). The *complement* of a subset A of $\mathfrak{A}$ is the set $A^* = \mathfrak{A} - A$. Demonstrate the following properties of complements for subsets of $\mathfrak{A}$:

$$(A^*)^* = A, \quad (A \cup B)^* = A^* \cap B^*, \quad \text{and} \quad (A \cap B)^* = A^* \cup B^*.$$

Show that $A^* * B^* = A * B$.

9.   The *cartesian product* of two sets S and T is the set $S \times T$ of ordered pairs (x, y) with $x \in S$ and $y \in T$. Two elements (x, y) and (x', y') of the cartesian product $S \times T$ are equal if and only if $x = x'$ and $y = y'$. Note that the cartesian product $T \times S$ is *not* the same as the cartesian product $S \times T$. (Why?)

As an example we may consider the coordinate plane as the set $\mathbf{R} \times \mathbf{R}$ where $\mathbf{R}$ denotes the set of real numbers. Each point of the coordinate plane is specified by an ordered pair (x, y) of real numbers, and each such ordered pair specifies a point in the plane. Note that $(x, y) = (y, x)$ if and only if $x = y$.

Let $[a, b] = \{x \in \mathbf{R} \mid a \leq x \leq b\}$ denote the closed interval from a to b. Then the cartesian product $[1, 3] \times [0, 1]$ may be represented in the coordinate plane by Figure 1.

Let $S_1, S_2, \ldots, S_n$ be sets. We define

$$\mathop{\Large\times}_{i=1}^{n} S_i = S_1 \times S_2 \times \cdots \times S_n$$

to be the set of ordered n-tuples $(x_1, x_2, \ldots, x_n)$ with $x_i \in S_i$. We shall call x_i the *i-th coordinate* of $(x_1, x_2, \ldots, x_n)$.

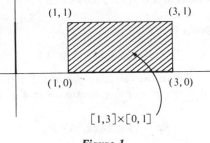

Figure 1

9α. Let A and C be subsets of S, and let B and D be subsets of T. Prove the following statements about subsets of $S \times T$:

$$(A \times B) \cap (C \times D) = (A \cap C) \times (B \cap D),$$
$$(A \cup C) \times (B \cup D) = (A \times B) \cup (A \times D) \cup (C \times B) \cup (C \times D),$$
$$(A \times B) - (C \times D) = (A \times (B - D)) \cup ((A - C) \times B).$$

9β. Let R, S, and T be sets. Are the sets $(R \times S) \times T$ and $R \times (S \times T)$ the same?

Mappings

---

10. Mapping is an abstraction of the concept of function. While a function assigns to a given number another number, a mapping assigns to a given element of one set an element of another. In other words, a mapping f from a set X to a set Y is a rule which assigns to each element $x \in X$ an element $y \in Y$. To remove the ambiguity residing in the word *rule*, it is necessary to recast this definition in the context of set theory.

A *mapping* f with *domain* X and *range* Y is a subset of $X \times Y$ such that for each element $x \in X$ there is precisely one element $y \in Y$ for which $(x, y) \in f$. We write $f: X \to Y$ to indicate that f is a mapping with domain X and range Y. If $f: X \to Y$ and $(x, y) \in f$, we usually write fx for y. It is now fashionable to write $f: x \mapsto y$ in place of $(x, y) \in f$.

Since mappings are defined as sets (of a special type), it is clear what equality of mappings should mean. Two mappings $f, g: X \to Y$ are equal if they are equal as subsets of $X \times Y$. It follows that $f = g$ if and only if $fx = gx$ for all $x \in X$.

The *identity mapping* of a set X is

$$1_X = \{(x, y) \in X \times X \mid x = y\}.$$

10α. An arbitrary subset of the cartesian product $X \times Y$ is called a *relation* with domain X and range Y. For any relation $R \subset X \times Y$ and any element $x \in X$, we set

$$Rx = \{y \in Y \mid (x, y) \in R\}.$$

A mapping is a special type of relation. Specifically, a relation $R \subset X \times Y$ is a mapping from X to Y if and only if for each $x \in X$, Rx consists of precisely one element of Y. Let **R** denote the set of all real numbers. Which of the following relations are mappings from **R** to **R**?

$$R_1 = \{(x, y) \in \mathbf{R} \times \mathbf{R} \mid x^2 + y^2 = 1\},$$
$$R_2 = \{(x, y) \in \mathbf{R} \times \mathbf{R} \mid xy = 1\},$$
$$R_3 = \{(x, y) \in \mathbf{R} \times \mathbf{R} \mid x^4 + y^3 = 1\},$$
$$R_4 = \{(x, y) \in \mathbf{R} \times \mathbf{R} \mid x^3 + y^4 = 1\},$$
$$R_5 = \{(x, y) \in \mathbf{R} \times \mathbf{R} \mid \sqrt{x} + \sqrt{y} = 1\}.$$

11. Let $f: X \to Y$ be a mapping. For any subset A of X, the *image of A by f* is the set

$$fA = \{y \in Y \mid y = fx, \, x \in A\}.$$

The set fX is also denoted $\operatorname{Im} f$ and called simply the *image of f*. The mapping f is called *onto* if $\operatorname{Im} f = Y$. If $\operatorname{Im} f$ is a proper subset of Y, then f is called *into*.

12. Let $f: X \to Y$ be a mapping. For any subset B of Y, the *inverse image of B by f* is the set

$$f^{-1}B = \{x \in X \mid fx \in B\}.$$

Note that $f^{-1}Y = X$. The mapping f is said to be *one to one* if for each $y \in Y$, $f^{-1}\{y\}$ has at most one element. ($f^{-1}\{y\} = \emptyset$ if $y \notin \operatorname{Im} f$.)

12α. Let $f: X \to Y$ be a mapping, let A and B be subsets of X, and let C and D be subsets of Y. Give a proof or counterexample for each of the following assertions:

$$f(A \cup B) = fA \cup fB, \qquad f^{-1}(C \cup D) = f^{-1}C \cup f^{-1}D,$$
$$f(A \cap B) = fA \cap fB, \qquad f^{-1}(C \cap D) = f^{-1}C \cap f^{-1}D,$$
$$f(A - B) = fA - fB, \qquad f^{-1}(C - D) = f^{-1}C - f^{-1}D,$$
$$f^{-1}(fA) = A, \qquad\qquad\quad f(f^{-1}C) = C.$$

Which of the false statements become true when f is one to one?

12β. For what integral values of n is the mapping $f: \mathbf{R} \to \mathbf{R}$ given by $f(0) = 0$ and $fx = x^n$ for $x \neq 0$, a one-to-one mapping?

13. A mapping $f: X \to Y$ is a *one-to-one correspondence* if f is one to one and onto. This is equivalent to saying that for each $y \in Y, f^{-1}\{y\}$ has precisely one element.

If $f: X \to Y$ is a one-to-one correspondence, we can define an inverse mapping $f^{-1}: Y \to X$ which is also a one-to-one correspondence. In fact, we just set

$$f^{-1} = \{(y, x) \in Y \times X \,|\, (x, y) \in f\}.$$

It follows that $y = fx$ if and only if $x = f^{-1}y$. Note that $(f^{-1})^{-1} = f$.

13α. Let $\mathbf{N}_k = \{1, 2, \ldots, k\}$. Define a one-to-one correspondence from $\mathbf{N}_k \times \mathbf{N}_l$ to $\mathbf{N}_{kl}$.

13β. If S and T denote sets, define a one-to-one correspondence from $S \times T$ to $T \times S$.

13γ. If R, S, and T denote sets, define a one-to-one correspondence from $(R \times S) \times T$ to $R \times (S \times T)$.

14. The *power set* of a set X is the set 2^X, whose elements are the subsets of X. In other words,

$$2^X = \{S \,|\, S \subset X\}.$$

Theorem. *There is no one-to-one correspondence $f: X \to 2^X$ for any set X.*

Proof. Suppose there were a set X with a one-to-one correspondence $f: X \to 2^X$. For each $x \in X$, fx is a subset of X and either $x \in fx$ or $x \notin fx$. Let

$$R = \{x \in X \,|\, x \notin fx\}.$$

Since a one-to-one correspondence is onto, $R = fa$ for some $a \in X$. Then $a \in fa = R$ implies $a \notin fa$, while $a \notin fa$ implies $a \in R = fa$. This is a contradiction.

14α. Construct a one-to-one correspondence

$$\phi: 2^{(A \cup B)} \to 2^A \times 2^B$$

where A and B are disjoint sets.

14β. Let $\mathbf{N}_k = \{1, 2, \ldots, k\}$. Construct a one-to-one correspondence between the sets $2^{\mathbf{N}_k}$ and $\mathbf{N}_{2^k} = \{1, 2, 3, \ldots, 2^k\}$.

15. Let **N** denote the set of *natural numbers* $\{1, 2, 3, \ldots\}$ and let $\mathbf{N}_k$ denote the subset $\{1, 2, \ldots, k\}$.

A set S is *finite* if it is in one-to-one correspondence with one of the sets $\mathbf{N}_k$, or if it is empty. If there is a one-to-one correspondence from a set S to $\mathbf{N}_k$, then the number of elements in S is k. A set which is not finite is called *infinite*. We cannot properly speak of the *number* of elements in an infinite set. However, we shall say that two infinite sets have the same *cardinality* if there is a one-to-one correspondence between them.

A set S is *countable* if it is in one-to-one correspondence with **N**, the set of natural numbers. For example, a one-to-one correspondence $f: \mathbf{N} \to \mathbf{Z}$ is given by

$$fk = (-1)^k [k/2],$$

where $[x]$ denotes the greatest integer not exceeding x. Consequently, the set **Z** of all integers is countable. A one-to-one correspondence $\phi: \mathbf{N} \times \mathbf{N} \to \mathbf{N}$ is given by

$$\phi(m, n) = \tfrac{1}{2}(m + n - 2)(m + n - 1) + n.$$

Consequently, $\mathbf{N} \times \mathbf{N}$ is a countable set.

Not every infinite set is countable: there can be no one-to-one correspondence between **N** and its power set $2^{\mathbf{N}}$; hence, the set $2^{\mathbf{N}}$ is *uncountable*. (We shall apply the words *countable* and *uncountable* to infinite sets only.)

15α. For any finite set S, let $\divideontimes S$ denote the number of elements of S. Prove that for any finite set S, $\divideontimes(2^S) = 2^{(\divideontimes S)}$.

15β. Prove that for any two finite sets S and T,

$$\divideontimes(S \cup T) + \divideontimes(S \cap T) = \divideontimes S + \divideontimes T.$$

15γ. Prove that for any two finite sets S and T,

$$\divideontimes(S \times T) = (\divideontimes S)(\divideontimes T).$$

15δ. Prove that the cartesian product of two countable sets is countable.

15ε. Prove that the set **Q** of all rational numbers is countable.

15ζ. Let X^{∞} denote the set of sequences of elements of X. Show that X^{∞} is uncountable if X has two or more elements. ("Sequence" here simply means an infinite string, $x_1, x_2, x_3, \ldots$, of elements of X.)

15η. Let S be a set with a countable number of elements. Show that a subset of S is either finite or countable.

150. Explain why a finite set cannot be in one-to-one correspondence with one of its proper subsets. (In some versions of set theory this is used as the defining property for finiteness.)

16. Let $f: X \to Y$ and $g: Y \to Z$ be mappings. Their *composite* is the mapping $gf: X \to Z$ determined by $(gf)x = g(fx)$. We may define gf more formally by

$$gf = \{(x, z) \in X \times Z \mid (fx, z) \in g\}.$$

If $f: X \to Y$, $g: Y \to Z$, and $h: Z \to W$ are mappings, then $h(gf) = (hg)f$; that is to say, *composition of mappings is associative*. To prove this, we merely need to observe that the two mappings $h(gf)$ and $(hg)f$ have the same value on each element $x \in X$:

$$(h(gf))x = h((gf)x) = h(g(fx)) = (hg)(fx) = ((hg)f)x.$$

If $f: X \to Y$ is a one-to-one correspondence with inverse $f^{-1}: Y \to X$, then $f^{-1}f = 1_X$ and $ff^{-1} = 1_Y$. Note that for any mapping $f: X \to Y$, we always have $f1_X = f = 1_Y f$.

Equivalence Relations

17. An *equivalence relation* on a set X is a subset R of $X \times X$ such that:

(a) $(x, x) \in R$ for all $x \in X$,
(b) $(x, y) \in R$ implies $(y, x) \in R$,
(c) $(x, y) \in R$ and $(y, z) \in R$ imply $(x, z) \in R$.

Frequently we prefer to write $x \equiv y \, (R)$ in place of $(x, y) \in R$.

If R is an equivalence relation on X, then for $x \in X$, the *R-equivalence class of x* is the set

$$[x]_R = \{y \in X \mid x \equiv y \, (R)\} = \{y \in X \mid (x, y) \in R\}.$$

When only one equivalence relation is under consideration, we usually suppress the subscript R on $[x]_R$.

For any equivalence relation R on X, $[x]_R = [y]_R$ if $x \equiv y \, (R)$, and $[x]_R \cap [y]_R = \emptyset$ if $x \not\equiv y \, (R)$. The *quotient of X by R* is the set X/R of equivalence classes $[x]_R$, where x runs through the elements of X. $[]_R: X \to X/R$ will denote the classifying map, defined by $[]_R x = [x]_R$.

17α. Let X be a set partitioned into disjoint subsets, $X_1, X_2, \ldots, X_n$. (Every element belongs to precisely one of the subsets.) Define an equivalence relation R on X for which $X/R = \{X_1, X_2, \ldots, X_n\}$.

17β. Prove that the intersection of equivalence relations is again an equivalence relation.

17γ. Let R be an equivalence relation on X and S an equivalence relation on X/R. Find an equivalence relation T on X such that $(X/R)/S$ is in one-to-one correspondence with X/T under the mapping $[[x]_R]_S \mapsto [x]_T$.

18. *Congruence of Integers.* Let m be a natural number, and let

$$R_m = \{(a, b) \in \mathbf{Z} \times \mathbf{Z} \mid a = b + km; k \in \mathbf{Z}\}.$$

R_m is an equivalence relation on the set $\mathbf{Z}$ of all integers and is called *congruence modulo m*. The number m is called the *modulus*. We write $a \equiv b \bmod m$ to indicate that $(a, b) \in R_m$; similarly, we write $a \not\equiv b \bmod m$ to indicate that $(a, b) \notin R_m$.

The equivalence class of $a \in \mathbf{Z}$ will be denoted $[a]_m$; that is to say,

$$[a]_m = \{x \in \mathbf{Z} \mid x \equiv a \bmod m\} = \{x \in \mathbf{Z} \mid x = a + km; k \in \mathbf{Z}\}.$$

Every $a \in \mathbf{Z}$ is congruent modulo m to one of the numbers $0, 1, \ldots, m - 1$. In fact if r is the smallest nonnegative integer in $[a]_m$, then $0 \leq r < m$ and $a \equiv r \bmod m$. It follows that the quotient set $\mathbf{Z}_m = \mathbf{Z}/R_m$ is simply

$$\{[0]_m, [1]_m, \ldots, [m - 1]_m\}.$$

18α. Prove that $a \equiv a' \bmod m$ and $b \equiv b' \bmod m$ imply that

$$a + b \equiv a' + b' \bmod m \quad \text{and} \quad ab \equiv a'b' \bmod m.$$

(This allows us to define sum and product on $\mathbf{Z}_m$ by the rules $[a]_m + [b]_m = [a + b]_m$ and $[a]_m[b]_m = [ab]_m$.)

18β. Let R_m denote congruence mod m on the set of integers $\mathbf{Z}$. What is the equivalence relation $R_m \cap R_n$?

18γ. Let

$$R = \{(a, b) \in \mathbf{Z} \times \mathbf{Z} \mid a^2 \equiv b^2 \bmod 7\}.$$

Into how many equivalence classes does R partition $\mathbf{Z}$?

19. Frequently, we are given a mapping $f: X \to Y$ and an equivalence relation R on the set X, and we want to define a mapping $\phi: X/R \to Y$ such that $\phi[\ \ ]_R = f$. Clearly, this can be done, if at all, only by setting $\phi[x]_R = fx$.

When does this make sense? If x and y are two elements of X which are equivalent with respect to the relation R, then we have $[x]_R = [y]_R$; it will have to follow that $\phi[x]_R = \phi[y]_R$, or what is the same thing, $fx = fy$. We see now that the formula $\phi[x]_R = fx$ defines a mapping $\phi \colon X/R \to Y$ if and only if for all $(x, y) \in R$, we have $fx = fy$. When this condition holds, we say that the mapping ϕ is *well defined*. (The terminology is idiotic: ϕ is not defined *at all* unless f has the required property.)

To give an example, suppose that we want to define a mapping from the set $\mathbf{Z}_m$ (defined in **18**) to any set Y by means of a mapping $f \colon \mathbf{Z} \to Y$. To do this, we must check that for any integers x and y, the condition $x \equiv y \bmod m$ implies $fx = fy$.

19α. When is the mapping $\phi \colon \mathbf{Z}_m \to \mathbf{Z}_n$ given by $\phi[x]_m = [x]_n$ well defined?

19β. Show that addition of elements of $\mathbf{Z}_m$ is well defined by the rule

$$[a]_m + [b]_m = [a + b]_m.$$

---

20. Let $\mathbf{N}$ denote the set of natural numbers $\{1, 2, 3, \ldots\}$. We shall take the following statement as an axiom:

Every nonempty subset of $\mathbf{N}$ has a smallest element.

This axiom has, as an immediate consequence, the *principle of mathematical induction*:

If S is a subset of $\mathbf{N}$ such that $1 \in S$ and such that $n \in S$ implies $n + 1 \in S$, then $S = \mathbf{N}$.

In fact the hypotheses on S imply that the set $\mathbf{N} - S$ has no smallest element.

20α. Prove the alternate form of the principle of mathematical induction: *If S is a subset of $\mathbf{N}$ such that $\mathbf{N}_1 \subset S$ and such that $\mathbf{N}_k \subset S$ implies $\mathbf{N}_{k+1} \subset S$, then $S = \mathbf{N}$.* (Recall that $\mathbf{N}_k = \{1, 2, \ldots, k\}$.)

20β. Prove by induction the formulas

$$\sum_{i=1}^{n} i = \tfrac{1}{2}n(n + 1) \quad \text{and} \quad \sum_{i=1}^{n} i^2 = \tfrac{1}{6}n(n + 1)(2n + 1).$$

20γ. Prove by induction the *binomial theorem*:

$$(x + y)^n = \sum_{k=0}^{n} \binom{n}{k} x^k y^{n-k},$$

where $\binom{n}{k}$ denotes the *binomial coefficient* $\dfrac{n!}{k!\,(n-k)!}$.

21. **The Division Theorem.** *For all natural numbers $a, b \in N$ there exist unique nonnegative integers q and r such that $a = qb + r$ and $r < b$.*

Proof. Let $\tilde{N}$ denote the set of nonnegative integers. It follows from **20** that every nonempty subset of $\tilde{N}$ has a smallest element. The set

$$S = \{x \in \tilde{N} \mid x = a - kb, k \in \tilde{N}\}$$

is not empty because $a \in S$. Let r be the smallest element of S. Clearly, $r = a - qb$ for some $q \in \tilde{N}$ and $r < b$ (otherwise $r - b \in S$, contradicting minimality of r). The uniqueness of r is apparent, and $r = a - qb = a - q'b$ implies $q' = q$, which shows uniqueness of q.

21α. Let $b \in N$, $b > 1$. Show that every natural number can be represented uniquely in the form

$$r_k b^k + r_{k-1} b^{k-1} + \cdots + r_1 b + r_0,$$

where $r_0, r_1, \ldots, r_k \in \{0, 1, \ldots, b - 1\}$.

22. A number $b \in N$ *divides* a number $a \in N$ provided $a = qb$ for some $q \in N$. To indicate that b divides a, we write $b \mid a$, and to indicate that it does not, $b \nmid a$. Thus, $2 \mid 4$, but $2 \nmid 5$. For any natural number n we always have $1 \mid n$ and $n \mid n$. If $n \neq 1$ and if 1 and n are the only natural numbers dividing n, then n is called a *prime number*. The first ten primes are 2, 3, 5, 7, 11, 13, 17, 19, 23, 29. The number of prime numbers is infinite (**22γ**).

22α. Show that every natural number other than 1 is divisible by some prime.

22β. Construct a natural number which is not divisible by any of the prime numbers in a given list of primes $p_1, p_2, \ldots, p_k$.

22γ. Prove that the number of primes is infinite.

23. If a and b are natural numbers, then among all the natural numbers dividing both a and b there is a largest one, which we call the *greatest common divisor* and denote by (a, b). For example, $(6, 8) = 2$, $(24, 30) = 6$, $(5, 7) = 1$. If $(a, b) = 1$, then we say that a and b are *relatively prime* or that a is *prime to b*.

Theorem. *If a and b are natural numbers, then there exist integers u and v such that $(a, b) = ua + vb$.*

Proof. Let

$$\mathfrak{A} = \{x \in \mathbf{N} \mid x = ma + nb \quad \text{for} \quad m, n \in \mathbf{Z}\}.$$

Since $a, b \in \mathbf{N}$ and $a = 1 \cdot a + 0 \cdot b$, $b = 0 \cdot a + 1 \cdot b$, the set $\mathfrak{A}$ is not empty and therefore has a smallest element $d = ua + vb$ for some $u, v \in \mathbf{Z}$. We claim that $d \mid x$ for all $x \in \mathfrak{A}$. Otherwise, for some $x \in \mathfrak{A}$ we have $x = qd + r$ where $0 < r < d$. Since $x = ma + nb$ for some $m, n \in \mathbf{Z}$, we have

$$r = x - qd = (ma + nb) - q(ua + vb) = (m - qu)a + (n - qv)b.$$

Consequently, $r \in \mathfrak{A}$ and $r < d$, which contradicts the choice of d as the smallest element of $\mathfrak{A}$. Thus, the claim is proved. It follows that $d \mid a$ and $d \mid b$, and therefore $1 \leq d \leq (a, b)$. However, $d = ua + vb$, and as a result we must have $(a, b) \mid d$, and hence $(a, b) \leq d$. It follows that $(a, b) = d = ua + vb$.

Corollary. *If p is a prime number and $p \mid ab$, then $p \mid a$ or $p \mid b$.*

Proof. Suppose $p \nmid a$. Then $(p, a) = 1 = ua + vp$ for some $u, v \in \mathbf{Z}$. Therefore $b = uab + vpb$, and $p \mid ab$ implies $p \mid b$.

23α. Prove that $d \in \mathbf{N}$ is the greatest common divisor of $a, b \in \mathbf{N}$ if and only if

(1) $d \mid a$ and $d \mid b$,
(2) $c \mid a$ and $c \mid b$ imply $c \mid d$.

23β. Prove that $m \equiv m' \bmod n$ implies $(m, n) = (m', n)$.

23γ. If a and b are natural numbers, then among all the natural numbers divisible by both a and b there is a smallest, which we call the *least common multiple* and denote by $[a, b]$. Show that $a \mid c$ and $b \mid c$ imply $[a, b] \mid c$. Show also that $(a, b)[a, b] = ab$.

23δ. Prove that $a \equiv b \bmod m$ and $a \equiv b \bmod n$ imply $a \equiv b \bmod [m, n]$.

23ε. Let $a, b, c \in \mathbf{N}$. Prove that

$$[a, (b, c)] = ([a, b], [a, c]) \quad \text{and} \quad (a, [b, c]) = [(a, b), (a, c)].$$

23ζ. *The Euclidean Algorithm.* Given $a, b \in \mathbf{N}$, define a decreasing sequence of natural numbers,

$$b = r_0 > r_1 > \cdots > r_n > r_{n+1} = 0$$

by the requirement that $r_{i-1} = q_i r_i + r_{i+1}$ for $i = 0, 1, \ldots, n$. (Let $r_{-1} = a$ so that $a = q_0 b + r_1$.) Show that $r_n = (a, b)$. (This method of computing the greatest common divisor is found at the beginning of the seventh book of Euclid's *Elements*.)

23η. Prove that if p is prime and $a, b \in \mathbf{Z}$,

$$(a + b)^p \equiv a^p + b^p \bmod p.$$

23θ. Let $a_1, a_2, \ldots, a_n \in \mathbf{Z}$, not all zero. Define the greatest common divisor $(a_1, a_2, \ldots, a_n)$ and prove the analogue of the theorem in **23**.

24.   *The Fundamental Theorem of Arithmetic.*   *Every natural number greater than 1 can be expressed uniquely as a product of prime numbers.*

Proof.   First we show that each $n \in \mathbf{N}$, $n > 1$, is divisible by some prime. Let S denote the set of natural numbers greater than 1 which are not divisible by any prime. If S is not empty, then S has a smallest element m. Since $m \mid m$, we cannot have m prime. Therefore, $m = ab$ where $1 < a < m$. Consequently, $a \notin S$ and there is a prime p which divides a. Then $p \mid m$ also and $m \notin S$, a contradiction. Therefore, S is empty.

Next we show that each $n \in \mathbf{N}$, $n > 1$, is a product of primes. Let S denote the set of natural numbers greater than 1 which cannot be written as a product of primes. If S is not empty, then S has a smallest element m and by the argument above, $m = pm'$ for some prime p. Since $m' < m$, we have $m' \notin S$. As a result m' can be written as a product of primes, $p_1 p_2 \cdots p_k$, or else $m' = 1$. Therefore, either $m = pp_1 p_2 \cdots p_k$ or $m = p$, and we have that m is a product of primes, which contradicts $m \in S$. Consequently, S is empty.

Finally, suppose there is a natural number greater than 1 which can be written in two ways as a product of primes:

$$n = p_1 p_2 \cdots p_k = q_1 q_2 \cdots q_l.$$

Then $p_1 \mid q_1 q_2 \cdots q_l$ and, by repeated use of the corollary of **23**, we may conclude that p_1 divides one of the q's, say $p_1 \mid q_1$. Since q_1 is prime, it follows that $p_1 = q_1$. As a result

$$p_2 p_3 \cdots p_k = q_2 q_3 \cdots q_l$$

and a similar argument shows that $p_2 = q_2$ (renumbering the q's if necessary). Continuing in the same manner, we arrive at the conclusion that $k = l$ and the two representations of n are identical (except for the order of the factors).

Corollary.   *Every natural number greater than 1 can be expressed uniquely in the form $p_1^{v_1} p_2^{v_2} \cdots p_k^{v_k}$ where $p_1, p_2, \ldots, p_k$ are prime numbers and $v_1, v_2, \ldots, v_k \in \mathbf{N}$.*

24α. Let $a = p_1^{v_1} p_2^{v_2} \cdots p_k^{v_k}$ and $b = p_1^{\mu_1} p_2^{\mu_2} \cdots p_k^{\mu_k}$. Show that

$$(a, b) = p_1^{\min(v_1, \mu_1)} p_2^{\min(v_2, \mu_2)} \cdots p_k^{\min(v_k, \mu_k)},$$

$$[a, b] = p_1^{\max(v_1, \mu_1)} p_2^{\max(v_2, \mu_2)} \cdots p_k^{\max(v_k, \mu_k)}.$$

24β. Compute the number of divisors of $n = p_1^{v_1} p_2^{v_2} \cdots p_k^{v_k}$.

25. *The Euler Function ϕ.* For any natural number n we let $\phi(n)$ denote the number of integers k such that $1 \le k \le n$ and $(k, n) = 1$. ϕ is called the *totient*, *indicator*, or *Euler ϕ-function*. Since the greatest common divisor (k, n) depends only upon the congruence class $[k]_n$ (23β), we may define $\phi(n)$ in another way as the number of elements in the set

$$\mathbf{Z}_n' = \{[k]_n \in \mathbf{Z}_n \mid (k, n) = 1\}.$$

Neither of these characterizations is useful in computing values of ϕ, but we shall use both to express $\phi(n)$ in terms of the unique factorization of n as a product of prime powers.

Proposition. *If p is prime, then $\phi(p^n) = p^n \left(1 - \dfrac{1}{p}\right)$*

Proof. Clearly, $(k, p^n) = 1$ if and only if $p \nmid k$. There are p^{n-1} numbers between 1 and p^n which are divisible by p, namely

$$1p, 2p, 3p, \ldots, (p^{n-1})p.$$

Therefore, $\phi(p^n) = p^n - p^{n-1} = p^n \left(1 - \dfrac{1}{p}\right)$.

Proposition. *If $(m, n) = 1$, then $\phi(mn) = \phi(m)\phi(n)$.*

Proof. We shall construct a one-to-one correspondence

$$\rho : \mathbf{Z}_{mn}' \to \mathbf{Z}_m' \times \mathbf{Z}_n'.$$

The proposition then follows immediately, because $\mathbf{Z}_{mn}'$ has $\phi(mn)$ elements and $\mathbf{Z}_m' \times \mathbf{Z}_n'$ has $\phi(m)\phi(n)$ elements. The mapping ρ is given by

$$\rho([k]_{mn}) = ([k]_m, [k]_n).$$

It is routine to verify that ρ is well defined.

The mapping ρ is one-to-one. Suppose $\rho([k]_{mn}) = \rho([k']_{mn})$. Then we have $[k]_m = [k']_m$ and $[k]_n = [k']_n$, or what is the same thing,

$$k \equiv k' \bmod m \quad \text{and} \quad k \equiv k' \bmod n.$$

Since $(m, n) = 1$, it follows that

$$k \equiv k' \bmod mn \quad \text{and} \quad [k]_{mn} = [k']_{mn}$$

(see **23δ**).

The mapping ρ is onto. Since $(m, n) = 1$, there are integers $u, v \in \mathbf{Z}$ such that $um + vn = 1$. Given $[a]_m \in \mathbf{Z}'_m$ and $[b]_n \in \mathbf{Z}'_n$, we set $k = bum + avn$. Then

$$k \equiv avn \equiv a \bmod m \quad \text{and} \quad k \equiv bum \equiv b \bmod n.$$

Hence, $[k]_m = [a]_m$ and $[k]_n = [b]_n$. What is more, $(k, mn) = 1$. If p is a prime and $p \mid mn$, then $p \mid m$ or $p \mid n$. If $p \mid m$, then $k \equiv a \bmod p$ and $a \not\equiv 0 \bmod p$ because $(a, m) = 1$; therefore $p \nmid k$. Similarly, $p \mid n$ implies $p \nmid k$. This shows that $p \mid mn$ implies $p \nmid k$ for any prime p, which implies that $(k, mn) = 1$.

Theorem.   *For every natural number*

$$\phi(n) = n\left(1 - \frac{1}{p_1}\right)\left(1 - \frac{1}{p_2}\right) \cdots \left(1 - \frac{1}{p_k}\right),$$

where $p_1, p_2, \ldots, p_k$ are the distinct primes dividing n.

Proof.   Write n in the form $p_1^{v_1} p_2^{v_2} \cdots p_k^{v_k}$ as guaranteed by the corollary in **24**. From the two propositions above it follows that

$$\phi(n) = \phi(p_1^{v_1})\phi(p_2^{v_2}) \cdots \phi(p_k^{v_k}) = p_1^{v_1}\left(1 - \frac{1}{p_1}\right)p_2^{v_2}\left(1 - \frac{1}{p_2}\right) \cdots p_k^{v_k}\left(1 - \frac{1}{p_k}\right),$$

and the formula of the theorem follows immediately.

25α.   Prove that $\sum_{d \mid n} \phi(d) = n$. $\sum_{d \mid n}$ denotes the sum over all the divisors of n. For example, 6 has the divisors 1, 2, 3, and 6, so that

$$\sum_{d \mid 6} \phi(d) = \phi(1) + \phi(2) + \phi(3) + \phi(6).$$

25β.   *The Möbius Function.*   For every natural number n we define a number $\mu(n)$ by the rules:

(1) $\mu(1) = 1$,
(2) $\mu(n) = 0$ if $p^2 \mid n$ for some prime p,
(3) $\mu(n) = (-1)^k$ if $n = p_1 p_2 \cdots p_k$ is a product of distinct primes.

Show that $(m, n) = 1$ implies $\mu(mn) = \mu(m)\mu(n)$ and that

$$\phi(n) = \sum_{d \mid n} \mu(d) \cdot (n/d).$$

Group Theory

The theory of groups is the proper place to begin the study of abstract algebra. Not only were groups the first algebraic structures to be characterized axiomatically and developed systematically from an abstract point of view, but more important, the concept of group structure is basic to the development of more complex abstractions such as rings and fields. Furthermore, group theory has an enormous number of applications to many diverse areas of mathematics and physics. Hardly any other area of mathematics can match the theory of groups in elegance and usefulness.

This chapter is an exposition of the elementary theory of groups with emphasis on groups of finite order. Three advanced topics (the Sylow theorems, the Jordan-Hölder theorem, and simplicity of alternating groups) are included for applications and depth.

The Definition of Group Structure

26. A *group* is a set G with an operation (called the *group product*) which associates to each ordered pair (a, b) of elements of G an element ab of G in such a way that:

17

(1) for any elements $a, b, c \in G$, $(ab)c = a(bc)$;

(2) there is a unique element $e \in G$ such that $ea = a = ae$ for any element $a \in G$;

(3) for each $a \in G$ there is an element $a^{-1} \in G$ such that $a^{-1}a = e = aa^{-1}$.

To be precise a group product should be viewed as a mapping $\mu: G \times G \to G$, and the group should be denoted (G, μ) to emphasize the role of the product. This notation would distinguish groups (G, μ) and (G, ν) which have the same underlying set but different products. However, such strict formalism obscures intuition and creates notational nuisances.

A set $G = \{e\}$ with the single element e and product defined by $ee = e$ satisfies (1), (2), and (3) trivially and is consequently called a *trivial group*.

26α. A *semigroup* is a set S with a product which associates to each ordered pair (a, b) of elements of S an element $ab \in S$ in such a way that $(ab)c = a(bc)$ for any elements $a, b, c \in S$. Show that the set of all mappings from a given set X to itself forms a semigroup in which the product is composition of mappings. Show that the set of all one-to-one correspondences of X with itself forms a group under composition.

26β. Let S be a semigroup with an element e such that $ea = a = ae$ for all $a \in S$. Show that e is unique. (This indicates that the word unique in (2) above is superfluous. It is used to insure the absolute clarity of (3).)

26γ. Let S be a semigroup with an element e such that $ea = a$ for all $a \in S$ and such that for every $a \in S$ there exists $a^{-1} \in S$ for which $a^{-1}a = e$. Prove that S is a group.

26δ. Let S be a semigroup with a finite number of elements. Suppose that the two cancellation laws hold in S; that is, if either $ab = ac$ or $ba = ca$, then $b = c$. Show that S is a group.

26ε. Let G be a group. Define a new product on G by $a * b = ba$ for any $a, b \in G$. Show that G^* (the set G with product $*$) is a group. G^* is called the *opposite group* to G.

26ζ. Let G and G' be groups. Define a product operation on the set $G \times G'$ by the rule $(a, a')(b, b') = (ab, a'b')$. Show that $G \times G'$ is a group under this product. ($G \times G'$ is called the *direct product* of G and G'.)

26η. A *symmetry* of a geometric figure is a one-to-one correspondence of the figure with itself preserving the distance between points; in other words, a symmetry is a self-congruence. The set of all symmetries of a given figure forms a group under composition. (Why?) For example, the group of symmetries of a line segment AB consists of two elements, the identity and the symmetry reversing A and B. Show that a symmetry of an equilateral triangle ABC is completely determined by the way it transforms the vertices. Make a complete list of the elements of the group of symmetries of ABC.

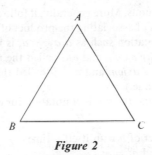

Figure 2

26θ. The group of symmetries of a regular polygon of n sides is called the *dihedral group* D_n. How many elements does D_n have?

26ι. Let V be a set with the four elements e, a, b, c on which a product is defined by the "multiplication table" in Table 3. Verify that V is a group. (V is known as the *four-group* or *Viergruppe* of Felix Klein.)

Table 3

	e	a	b	c
e	e	a	b	c
a	a	e	c	b
b	b	c	e	a
c	c	b	a	e

26κ. Show that the power set 2^X of any set X is a group under the operation of symmetric difference $A * B$ (8δ).

26λ. Show that the set $(-1, 1)$ of real numbers x such that $-1 < x < 1$ forms a group under the operation $x \cdot y = (x + y)/(1 + xy)$.

26μ. Find an operation on the set $(0, 1)$ of real numbers $x, 0 < x < 1$, which makes $(0, 1)$ a group in such a way that the inverse of x is $1 - x$.

26ν. Generalize the definition of direct product given in **26ζ** to obtain a definition of the direct product $G_1 \times G_2 \times \cdots \times G_n$ of n groups, $G_1, G_2, \ldots, G_n$.

27. Statements (1), (2), and (3) of **26** are known as the *axioms of group structure*. Group structure may be axiomatically characterized in several ways, but the particular way given here is the most direct and convenient.

(1) is called the *associativity axiom*, because it states that the two ways of associating a product of three elements are equal. Consequently, the

notation *abc* is unambiguous. More generally, it follows (after some argument) that all the various ways of associating the product of any number of elements are equal. Therefore, notation such as $a_1 a_2 \cdots a_n$ is unambiguous.

(2) is called the *identity axiom* and *e* is called the *identity element*.

(3) is called the *inverse axiom* and a^{-1} is called the *inverse* of *a*. (In **28** we shall see that a^{-1} is unique.)

It is customary to extend the product notation for elements in the following ways:

a^n denotes the product of *a* with itself *n* times;

a^{-n} denotes $(a^{-1})^n$ and $a^0 = e$;

if *A* and *B* are subsets of the group *G*, then

$$AB = \{x \in G \mid x = ab, a \in A, b \in B\}.$$

27α.   Show that the five distinct ways of associating a product of four group elements in a given order are all equal.

27β.   Let $a_1 a_2 \cdots a_n$ be defined inductively by the rule:

$$a_1 a_2 \cdots a_n = (a_1 a_2 \cdots a_{n-1})a_n.$$

(This gives a particular association for the product of $a_1, a_2, \ldots, a_n$.) Prove that

$$(a_1 a_2 \cdots a_n)(a_{n+1}a_{n+2} \cdots a_{n+m}) = a_1 a_2 \cdots a_{n+m}.$$

27γ.   With the result of **27β** prove the *general associative law*, that all the ways of associating a product of any number of elements in a given order are equal. (This means that expressions such as $a_1 a_2 \cdots a_n$ are unambiguous.)

28.   Proposition.   *For any elements a, b, c, d of a group, it is true that*
  (1) $ab = e$ *implies* $b = a^{-1}$;
  (2) $(c^{-1})^{-1} = c$;
  (3) $(cd)^{-1} = d^{-1}c^{-1}$.
(Note that (1) implies that inverses are unique.)

Proof.

  (1) $b = eb = (a^{-1}a)b = a^{-1}(ab) = a^{-1}e = a^{-1}$.
  (2)   Apply (1) with $a = c^{-1}$ and $b = c$.
  (3)   Set $a = cd$ and $b = d^{-1}c^{-1}$.   Then   we   have $ab = (cd)(d^{-1}c^{-1}) = c(dd^{-1})c^{-1} = cc^{-1} = e$ and by (1) it follows that $d^{-1}c^{-1} = b = a^{-1} = (cd)^{-1}$.

28α.   Prove that for any elements $a_1, a_2, \ldots, a_n$ of a group,

$$(a_1 a_2 \cdots a_n)^{-1} = a_n^{-1} \cdots a_2^{-1}a_1^{-1}.$$

28β.   Let a be an element of a group G. Show that the mapping $\lambda_a : G \to G$ given by $\lambda_a\, g = ag$ for any $g \in G$ is a one-to-one correspondence.

28γ.   A group G is *isomorphic* to a group G' if there exists a one-to-one correspondence $\phi : G \to G'$ such that $\phi(ab) = (\phi a)(\phi b)$, or in other words, such that ϕ preserves group products. Show that such a mapping ϕ (called an *isomorphism*) also preserves identity elements and inverses.

28δ.   Let S be a set with an operation which assigns to each ordered pair (a, b) of elements of S an element a/b of S in such a way that:
   (1) there is an element $1 \in S$, such that $a/b = 1$ if and only if $a = b$;
   (2) for any elements $a, b, c \in S$, $(a/c)/(b/c) = a/b$.
Show that S is a group under the product defined by $ab = a/(1/b)$.

29.   A group G is *abelian* if $ab = ba$ for all $a, b \in G$. Abelian groups are named for Neils Henrik Abel (1802–1829), who discovered their importance in his research on the theory of equations. It is often convenient to use additive notation for abelian groups: the group product of two elements a and b is written $a + b$; 0 denotes the identity element; $-a$ denotes the inverse of a. Then the axioms of group structure read:

   (1) $(a + b) + c = a + (b + c)$,
   (2) $0 + a = a = a + 0$,
   (3) $-a + a = 0 = a + (-a)$,
and of course the defining property of abelian group structure,
   (4) $a + b = b + a$.

29α.   Which of the following sets are abelian groups under the indicated operations?

   (1) The set $\mathbf{Q}$ of rational numbers under addition; under multiplication.
   (2) The set $\mathbf{Q}^*$ of nonzero rational numbers under addition; under multiplication.
   (3) The set $\mathbf{Q}^+$ of positive rational numbers under addition; under multiplication.
   (4) The set $\mathbf{Q}^{2 \times 2}$ of 2×2 matrices with rational entries under addition; under multiplication.
   (5) The set $\mathbf{Q}_1^{2 \times 2}$ of 2×2 matrices with determinant 1 under addition; under multiplication.

29β.   Prove that a group with less than six elements is abelian. Construct a group with six elements which is not abelian, and one which is.

29γ.   Prove that for $n > 2$ the dihedral group D_n is non-abelian.

29δ.   Show that a group G in which $x^2 = e$ for every $x \in G$, is an abelian group.

Examples of Group Structure

---

30. *The Symmetric Group on n Letters.* Let N_n denote the set $\{1, 2, \ldots, n\}$. A permutation of n letters is a one-to-one, onto mapping from N_n to N_n. S_n will denote the set of all permutations of n letters. If $\pi, \rho \in S_n$, then clearly the composite mapping $\pi\rho \in S_n$. The operation which assigns to each ordered pair (π, ρ) of permutations of n letters their composite, $\pi\rho$, is a group product on S_n:

(1) composition of mappings is associative as we have observed in **(16)**;
(2) the identity mapping **(10)** $\iota_n: N_n \to N_n$ is the identity element;
(3) for any $\pi \in S_n$, the inverse mapping **(13)** $\pi^{-1} \in S_n$ serves as an inverse element for π in the group-theoretic sense.

The group S_n is called the *symmetric group on n letters*. We shall devote considerable space to the theory of symmetric groups at the end of this chapter **(76–86)**.

30α. Show that the group S_n has $n!$ elements.

30β. Show that S_n is not abelian for $n > 2$.

30γ. Construct an isomorphism of the symmetric group S_3 with the dihedral group D_3. (See **26θ**.)

31. *The Circle Group.* Let K denote the set of points of unit distance from the origin in the plane of complex numbers $\mathbf{C}$. In other words

$$K = \{z \in \mathbf{C} \mid |z| = 1\}.$$

The set K is closed under the usual product of complex numbers: if $z, w \in K$, then $|zw| = |z|\,|w| = 1$ so that $zw \in K$. In fact K is a group under this multiplication:

(1) multiplication of complex numbers is associative;
(2) the identity element of K is the complex number $1 = 1 + 0i$;
(3) the inverse of $z \in K$ is just $1/z$.

K is called the *circle group* since the elements of K form a circle in the complex plane. Note that K is abelian.

32. *The Additive Group of Integers.* Addition is a group product on the set of integers $\mathbf{Z}$. The identity element is 0 and the inverse of n is $-n$. Of course $\mathbf{Z}$ is abelian.

33. *The Additive Group of Integers Modulo n.* Let Z_n denote the set of equivalence or congruence classes modulo the natural number n (**18**). Recall that we may take

$$Z_n = \{[0]_n, [1]_n, \ldots, [n-1]_n\}.$$

A group product on Z_n is given by addition of congruence classes: $[a]_n + [b]_n = [a+b]_n$. Of course we must verify that this addition is *well defined*; that is, we must show that $[a']_n = [a]_n$ and $[b']_n = [b]_n$ imply $[a'+b']_n = [a+b]_n$. (We leave it to the reader.) The identity element of Z_n is $[0]_n$ and the inverse of $[k]_n$ is

$$-[k]_n = [-k]_n = [n-k]_n.$$

As the additive notation suggests, Z_n is abelian.

34. *The Multiplicative Group Modulo n.* We can multiply elements of the set Z_n by setting $[a]_n[b]_n = [ab]_n$. (Check that this is well defined.) Clearly, for any integer k,

$$[k]_n[1]_n = [k]_n = [1]_n[k]_n.$$

Furthermore, $[1]_n$ is the only element of Z_n with this property since $[x]_n[k]_n = [k]_n$ implies that $xk \equiv k \bmod n$; taking $k = 1$, we obtain $x \equiv 1 \bmod n$, or what is the same thing, $[x]_n = [1]_n$. Consequently, $[1]_n$ is an identity element.

However, *this multiplication is not a group product for* Z_n, because some elements ($[0]_n$ for example) do not have an inverse. In fact $[k]_n \in Z_n$ has an inverse if and only if k and n are relatively prime. We show this as follows. Suppose $(k, n) = 1$. By **23** there are integers u and v such that $uk + vn = 1$. Thus,

$$[uk + vn]_n = [uk]_n = [u]_n[k]_n = [1]_n,$$

and $[u]_n$ is an inverse for $[k]_n$. On the other hand, if

$$[u]_n[k]_n = [uk]_n = [1]_n$$

for some integer u, then $uk \equiv 1 \bmod n$ and $uk + vn = 1$ for some integer v. This implies that $(k, n) = 1$. It is now a routine matter to verify that

$$Z_n' = \{[k]_n \in Z_n \mid (k, n) = 1\}$$

forms a group under multiplication of equivalence classes. Note that for a prime number p we have

$$Z_p' = \{[1]_p, [2]_p, \ldots, [p-1]_p\}.$$

We shall call Z_n' the *multiplicative group modulo n*.

Subgroups and Cosets

---

35. A *subgroup* of a group G is a nonempty subset H such that (1) $a, b \in H$ implies $ab \in H$, and (2) $a \in H$ implies $a^{-1} \in H$.

Clearly a subgroup H of a group G is a group in its own right under the group product inherited from G. The sets $\{e\}$ and G are subgroups of a group G. A subgroup H of a group G is a *proper subgroup* when H is a proper subset of G. The subgroup $\{e\}$ is *trivial*; all others are called *nontrivial*.

Proposition. *If H is a finite subset of a group G and $a, b \in H$ implies $ab \in H$, then H is a subgroup of G.*

Proof. We need only show that $a \in H$ implies $a^{-1} \in H$. A simple induction argument shows that

$$\{x \in G \mid x = a^n; n \in \mathbf{N}\} \subset H$$

whenever $a \in H$. Since H is finite, it must happen that $a^n = a^m$ for some natural numbers $n > m$. Since $a^{-m} \in G$, we have $a^{n-m} = a^n a^{-m} = a^m a^{-m} = e$. Either $n = m + 1$ and $a = a^{n-m} = e$, or $n > m + 1$ and $a^k = a^{n-m} = e$ for $k > 1$. Thus, either $a^{-1} = e = a$, or $a^{-1} = a^{k-1}$. In either case $a^{-1} \in H$.

35α. Show that a nonempty subset H of a group G is a subgroup of G if and only if $a, b \in H$ implies $ab^{-1} \in H$.

35β. Let $H_1, H_2, \ldots, H_n$ be subgroups of a group G. Show that $H = \bigcap_{k=1}^{n} H_k$ is a subgroup of G.

35γ. Let H be a subgroup of G and let $a \in G$. Let

$$H^a = \{x \in G \mid axa^{-1} \in H\}.$$

Show that H^a is a subgroup of G. (H^a is called the *conjugate* of H by a.)
Let

$$N(H) = \{a \in G \mid H^a = H\}.$$

Show that $N(H)$ is a subgroup of G and that H is a subgroup of $N(H)$. ($N(H)$ is called the *normalizer* of H.)

35δ. Let Z_G denote the set of elements of a group G which commute with all the elements of G; that is,

$$Z_G = \{x \in G \mid xa = ax, a \in G\}.$$

Show that Z_G is a subgroup of G. (Z_G is called the *center* of G.)

35ε. If S is any subset of a group G, then there is a smallest subgroup of G, say H, containing S. Why? In this case we say that S *generates* H. Show that S generates H if and only if every element of H can be written as a product $s_1 s_2 \cdots s_m$ where $s_i \in S$ or $s_i^{-1} \in S$ for each i. (When S generates H, we write $H = \langle S \rangle$.)

35ζ. The dihedral group D_6 **(26θ)** is the group of symmetries of a regular hexagon, $ABCDEF$. Let $\alpha \in D_6$ denote rotation counterclockwise by $60°$, and β denote reflection in the horizontal AD axis. Then D_6 consists of the twelve

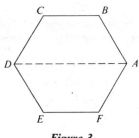

Figure 3

elements $\alpha^i \beta^j$ where i runs through 0, 1, 2, 3, 4, 5 and $j = 0, 1$. Table 4 indicates how each element of D_6 acts on the vertices of $ABCDEF$.

Table 4

e	α	α^2	α^3	α^4	α^5	β	$\alpha\beta$	$\alpha^2\beta$	$\alpha^3\beta$	$\alpha^4\beta$	$\alpha^5\beta$
A	B	C	D	E	F	A	B	C	D	E	F
B	C	D	E	F	A	F	A	B	C	D	E
C	D	E	F	A	B	E	F	A	B	C	D
D	E	F	A	B	C	D	E	F	A	B	C
E	F	A	B	C	D	C	D	E	F	A	B
F	A	B	C	D	E	B	C	D	E	F	A

(1) Determine the subgroup of D_6 leaving C fixed.
(2) Determine the subgroup in which A, C, and E are permuted and show that it is isomorphic to D_3.

    (3) Determine the subgroup generated by α^4 and $\alpha^3\beta$.

    (4) Show how each of the elements $\beta\alpha^k$ for $k = 1, 2, 3, 4, 5$ may be written in the form $\alpha^i\beta^j$.

    (5) Determine the center of D_6. (35δ.)

35η. Show that the dihedral group D_n is generated by two elements α and β satisfying $\alpha^n = e$, $\beta^2 = e$, and $\beta\alpha = \alpha^{n-1}\beta$.

35θ. Let X, Y, and Z be subgroups of a group G. Show that $Y \subset X$ implies $X \cap YZ = Y(X \cap Z)$.

36. *Proposition.* *Every nontrivial subgroup of* $\mathbf{Z}$, *the additive group of integers, has the form* $n\mathbf{Z} = \{x \in \mathbf{Z} \mid n \mid x\}$ *for some natural number* n.

Proof. A nontrivial subgroup H of $\mathbf{Z}$ contains some integer $m \neq 0$. Since H also contains $-m$, the set $H \cap \mathbf{N}$ is not empty and therefore has a least element n. It clearly follows that $an \in H$ for every $a \in \mathbf{Z}$, or in other words, $n\mathbf{Z} \subset H$. Suppose $m \in H - n\mathbf{Z}$. Then $m \neq 0$ and we may assume $m > 0$. (Otherwise replace m by $-m$.) By the division theorem (21) $m = qn + r$, and we have $0 < r < n$, since $r = 0$ implies $m = qn \in n\mathbf{Z}$. Now we have $r = m - qn \in H$ and $0 < r < n$, which imply that n is not the least element of $H \cap \mathbf{N}$. This contradiction forces the conclusion that $H - n\mathbf{Z} = \emptyset$ or $H = n\mathbf{Z}$.

36α. Describe the subgroup $n\mathbf{Z} \cap m\mathbf{Z}$ of $\mathbf{Z}$.

36β. Describe the subgroup of $\mathbf{Z}$ generated by n and m, that is, by the subset $\{n, m\}$.

36γ. Describe all the subgroups of $\mathbf{Z}_n$.

36δ. Show that $\mathbf{Z} \times \mathbf{Z}$ has subgroups not of the form $n\mathbf{Z} \times m\mathbf{Z}$.

37. *Congruence modulo a subgroup.* Let H be a subgroup of a group G. We use H to define an equivalence relation on G. Let

$$R_H = \{(x, y) \in G \times G \mid x^{-1}y \in H\}.$$

Certainly $(x, x) \in R_H$ for all $x \in G$, because $x^{-1}x = e \in H$. If $(x, y) \in R_H$, then $x^{-1}y \in H$, and consequently $(x^{-1}y)^{-1} \in H$. However, $(x^{-1}y)^{-1} = y^{-1}x$, and thus $(y, x) \in R_H$. Finally, (x, y) and (y, z) in R_H imply $x^{-1}y \in H$ and $y^{-1}z \in H$, which shows $(x^{-1}y)(y^{-1}z) = x^{-1}z \in H$, or $(x, z) \in R_H$. We have verified that R_H is an equivalence relation (17).

    We write $x \equiv y \bmod H$ when $(x, y) \in R_H$. Clearly, $x \equiv y \bmod H$ if and only if $y = xh$ for some $h \in H$. Consequently, we denote the equivalence class of $x \in G$ by xH to indicate that it consists of all the elements xh where $h \in H$. We call xH the *left coset of* x *modulo* H. The set of all equivalence classes (left cosets), denoted by G/H, is called the *left coset space of G modulo H*.

Since left cosets are equivalence classes, *two left cosets are either identical or disjoint*. To rephrase, $xH = yH$ when $x \equiv y \bmod H$, and $xH \cap yH = \emptyset$ when $x \not\equiv y \bmod H$. Note that $eH = H$.

An example of congruence modulo a subgroup is furnished by **Z** modulo $m\mathbf{Z}$. This is just congruence modulo the integer m as defined previously (**18**), and the quotient set $\mathbf{Z}/m\mathbf{Z}$ is just $\mathbf{Z}_m$. The coset of the integer k is denoted by $k + m\mathbf{Z}$ to conform with the additive notation for abelian groups, or $[k]_m$ as before.

37α. Show that all the left cosets of a group G with respect to a subgroup H have the same number of elements; in other words, show that any two left cosets are in one-to-one correspondence.

37β. Let H be a subgroup of a group G. Define an equivalence relation on G which partitions G into right cosets, that is, subsets of the form

$$Hy = \{x \in G \mid x = hy, h \in H\}.$$

Prove that the number of right cosets is the same as the number of left cosets of G with respect to H; that is, show that the set of right cosets and the set of left cosets are in one-to-one correspondence.

37γ. Show that when G is abelian, every right coset is a left coset modulo H.

37δ. Let an equivalence relation on S_n be defined by $\pi \sim \tau$ if and only if $\pi n = \tau n$. Show that this equivalence relation is congruence modulo a subgroup of S_n.

38. The *order o(G)* of a group G with a finite number of elements is just the number of elements of G. Thus, $o(\mathbf{Z}_n) = n$, $o(\mathbf{Z}_n') = \phi(n)$ (see **25, 34**). The order of the symmetric group S_n is the number of permutations of n letters (**30**). For any such permutation $\pi: \mathbf{N}_n \to \mathbf{N}_n$ there are n choices for $\pi(1)$, $n - 1$ choices for $\pi(2)$, and so forth; in all there are $n(n - 1) \cdots (1) = n!$ permutations of n letters. In other words $o(S_n) = n!$.

If a group has an infinite number of elements, it has *infinite order*. For example, the circle group K (**31**) and the additive group of integers **Z** (**32**) are groups of infinite order.

For brevity, a group of finite [infinite] order is called a *finite [infinite] group*.

38α. Let H and K be subgroups of a group G. Show that HK is a subgroup of G if and only if $HK = KH$. Show that when HK is a finite subgroup of G,

$$o(HK) = o(H)o(K)/o(H \cap K).$$

38β. Determine all the subgroups of the dihedral group D_6 and the order of each. (See **35ζ**.)

38γ. Let G be a nontrivial group with no proper subgroups except the trivial one. Show that G is finite and that the order of G is prime.

38δ. Prove that a group G of even order contains an element $a \neq e$ such that $a^2 = e$.

39. The *index* $[G: H]$ of a subgroup H of a group G is the number of left cosets of G modulo H, or in other words, the number of elements of the left coset space G/H, *provided this number is finite*. Otherwise, the index is said to be infinite.

For example, the left cosets of $\mathbf{Z}$ modulo $m\mathbf{Z}$ are the sets $m\mathbf{Z} + k$ where k runs through $0, 1, \ldots, m - 1$. Thus, $[\mathbf{Z}: m\mathbf{Z}] = m$.

Proposition. *If K is a subgroup of H and H a subgroup of G, then $[G: K] = [G: H][H: K]$, provided these indices are finite.*

Proof. It is easy to see that each coset gH of G modulo H contains the cosets $g(hK) = (gh)K$ of G modulo K, where hK runs through all the cosets of H modulo K.

39α. Let H denote the subgroup of S_n consisting of all elements $\pi \in S_n$ such that $\pi n = n$. What is $[S_n: H]$?

39β. If H and K are subgroups of finite index of a group G, show that $H \cap K$ is a subgroup of finite index of G and that

$$[G: H \cap K] \leq [G: H][G: K].$$

40.   *Lagrange's Theorem.* *If H is a subgroup of G, a group of finite order, then $[G: H] = o(G)/o(H)$.*

Proof. The map $f_y: H \to yH$ given by $f_y(h) = yh$ is a one-to-one correspondence. The inverse map $f_y^{-1}: yH \to H$ is given by $f_y^{-1}(x) = y^{-1}x$. Thus, the left coset yH has the same number of elements as H, namely $o(H)$. In fact each left coset of G has $o(H)$ elements. Since left cosets are identical or disjoint, each element of G belongs to precisely one left coset. There are $[G: H]$ left cosets, and therefore, $o(G) = [G: H]o(H)$. All three numbers are finite. Thus, $[G: H] = o(G)/o(H)$.

Lagrange's theorem is frequently stated as *the order of a subgroup of a group of finite order divides the order of the group.* Of course the equation $[G: H] = o(G)/o(H)$ makes sense only when $o(G)$ is finite (and consequently $o(H)$ and $[G: H]$ are also finite). However, even when $o(G)$ is infinite, a sensible interpretation can be made: if $[G: H]$ is finite, then $o(H)$ is infinite; if $o(H)$ is finite, then $[G: H]$ is infinite. In other words, *a subgroup of finite index in an infinite group is of infinite order; a finite subgroup of an infinite group has infinite index.*

Joseph Louis Lagrange (1736–1813) made contributions to number theory, the theory of equations, and the calculus of variations as well as other fields. His observations on the theory of equations set the stage for the development of group theory.

41. If a is an element of a group G, then the *order* of a, $o(a)$, is the order of the subgroup

$$\langle a \rangle = \{x \in G \mid x = a^n; n \in \mathbf{Z}\},$$

which is the smallest subgroup of G containing the element a. If $\langle a \rangle$ is finite, it happens that $a^n = a^m$ for some integers n and m such that $n > m$. Then, $a^{n-m} = e$, and $n - m > 0$. Let k be the smallest positive integer such that $a^k = e$. Writing any given integer n uniquely in the form $n = qk + r$, where $0 \le r < k$, we have

$$a^n = a^{qk+r} = (a^k)^q a^r = e^q a^r = a^r.$$

Thus, each a^n equals one of the elements $a^0 = e$, $a^1 = a$, $a^2, \ldots, a^{k-1}$. In other words,

$$\langle a \rangle = \{e, a, a^2, \ldots, a^{k-1}\},$$

and the order of a is $o(\langle a \rangle) = k$. Therefore, we can state: *the order of an element a of a group G is the smallest positive integer k such that $a^k = e$* (if such a k exists). If $\langle a \rangle$ is infinite, then for all positive k, $a^k \ne e$, and a is called an element of infinite order.

If a is an element of a finite group G, then LaGrange's theorem **(40)** implies that $o(a) \mid o(G)$, and it follows that $a^{o(G)} = e$. In other words, *the order of an element of a finite group divides the order of the group.* We use this principle in the next article to prove Euler's theorem.

41α. Determine the order of a^m where a is an element of order n in a group G.

41β. Let a and b be elements of an abelian group G. Describe $o(ab)$ in terms of $o(a)$ and $o(b)$.

41γ. Show that $o(axa^{-1}) = o(x)$ for any elements a and x of a group G.

41δ. For any elements a and b of a group G, show that $o(ab) = o(ba)$.

41ε. Let G be an abelian group. Show that for any natural number n, the set $G_n = \{x \in G \mid o(x) \mid n\}$ is a subgroup of G.

41ζ. Prove there can be only two distinct groups with order 4 (up to isomorphism). Do the same for order 6.

41η. Prove that a group with only a finite number of subgroups must be finite.

42.   Euler's Theorem.   *If a is any integer prime to m, then $a^{\phi(m)} \equiv 1 \bmod m$.*

Proof. Since $(a, m) = 1$, the congruence class modulo m of a belongs to Z'_m, the multiplicative group modulo m (**34**). By (**41**) the order k of $[a] \in Z'_m$ divides the order of Z'_m, which is just $\phi(m)$ where ϕ is the Euler function (**25**). Then, $[a]^k = [1]$ implies that $[a]^{\phi(m)} = [a^{\phi(m)}] = [1]$, or what is the same thing, $a^{\phi(m)} \equiv 1 \bmod m$.

Corollary.   *If p is a prime number, then $a^{p-1} \equiv 1 \bmod p$ for any integer a not divisible by p.*

The corollary is known as *Fermat's little theorem.*

43. A group G is *cyclic* if every element of G is a power a^k (k positive, negative, or zero) of a fixed element $a \in G$. The element a is said to *generate* the group G. *A cyclic group is always abelian.*

A cyclic group may be finite or infinite. If G is an infinite cyclic group generated by $a \in G$, then a is an element of infinite order in G and all the powers of a are distinct. Thus, $G = \{\ldots, a^{-2}, a^{-1}, e, a^1, a^2, \ldots\}$. The additive group of integers Z is an infinite cyclic group generated by the element $1 \in Z$. (Powers must be interpreted additively: $n \in Z$ is the n-th "power" of 1.)

If G is a cyclic group of finite order n generated by a, then $G = \{e, a, a^2, \ldots, a^{n-1}\}$ and the order of a is also n. The additive group of integers modulo m, Z_m, is a cyclic group of order m generated by $[1]_m \in Z_m$.

Proposition.   *A subgroup of a cyclic group is again cyclic.*

Proof. Let G be a cyclic group, generated by $a \in G$, with a subgroup H. We may assume $H \neq \{e\}$: a trivial group is always cyclic. Let k be the smallest element of the set $S = \{n \in N \mid a^n \in H\}$. Then if $a^n \in H$, we have $k \mid n$; otherwise $n = qk + r$ with $0 < r < k$, and $a^r = a^n(a^k)^{-q} \in H$ contradicting the choice of k. Thus, $a^n \in H$ implies $a^n = (a^k)^q$ where $q = n/k$. Consequently H is a cyclic group generated by a^k.

Proposition.   *A group of prime order is cyclic.*

Proof. If G has order p, where p is a prime number, then the order of any element a is 1 or p—the only divisors of p. Since the only element of order 1 is the identity e, any element $a \neq e$ has order p and generates G.

43α. What is the number of elements of order d in a cyclic group of order n? (See **25α.**)

43β. Let G be a finite group with the property that for any two subgroups H and K, either $H \subset K$ or $K \subset H$. Show that G is a cyclic group whose order is a power of a prime.

43γ. Let G and G' be cyclic groups the orders of which are relatively prime. Show that their direct product $G \times G'$ is cyclic.

43δ. Show that the direct product of two infinite cyclic groups is not cyclic.

44. *The n-th roots of unity.* As another example of a finite cyclic group we take up the group of n-th roots of unity. This group will play an important role later in the study of Galois theory.

A complex number $z \in \mathbf{C}$ is an n-th *root of unity* if $z^n = 1$. It follows from $z^n = 1$, that $|z|^n = 1$, $|z| = 1$, and $z = e^{i\theta}$ for some $\theta \in [0, 2\pi]$. Thus, every n-th root of unity is an element of the circle group K (**31**). Substituting $z = e^{i\theta}$ in $z^n = 1$, we have $e^{in\theta} = 1$, so that $n\theta$ must be multiple of 2π, or in other words, $\theta = 2\pi k/n$ for some integer k. The value of $e^{2\pi ik/n}$ depends only upon the congruence class of k modulo n. Therefore, there are precisely n n-th roots of unity, namely the complex numbers $e^{2\pi ik/n}$ for $k = 0, 1, 2, \ldots,$ $n - 1$. Letting ζ denote $e^{2\pi i/n}$, we see that $e^{2\pi ik/n} = \zeta^k$, and

$$K_n = \{1, \zeta, \zeta^2, \ldots, \zeta^{n-1}\}$$

is a complete set of n-th roots of unity. It is easily verified that K_n is a cyclic group of order n generated by ζ. Note that

$$\zeta^{qn+r} = (\zeta^n)^q \zeta^r = (1)^q \zeta^r = \zeta^r.$$

Now ζ is not the only element of K_n which generates K_n. (In general the generator of a cyclic group is not unique.) An element of K_n which generates K_n is called a *primitive n-th root of unity*. The following proposition identifies the primitive n-th roots of unity.

Proposition. $\zeta^k \in K_n = \{1, \zeta, \zeta^2, \ldots, \zeta^{n-1}\}$ *is a primitive n-th root of unity if and only if k is prime to n.*

Proof. If $(k, n) = 1$, then $1 = uk + vn$ for some $u, v \in \mathbf{Z}$ and

$$\zeta^m = \zeta^{muk + mvn} = \zeta^{muk} = (\zeta^k)^{mu}$$

for all m. Thus ζ^k generates K_n. On the other hand, if ζ^k generates K_n, we have $\zeta = (\zeta^k)^u$ so that $uk \equiv 1 \bmod n$ or $1 = uk + vn$ for some $u, v \in \mathbf{Z}$ and $(k, n) = 1$.

44α. Let C be a cyclic group with generator c. What other elements of C generate C? (Include the case in which C is infinite.)

44β. Let ζ be a primitive n-th root of unity, and let ξ be a primitive m-th root of unity. What is the subgroup of the circle group K generated by ζ and ξ?

44γ. Describe the group $K_n \cap K_m$.

Conjugacy, Normal Subgroups, and Quotient Groups

---

45. Let S be any subset of a group G, and let a be any element of G. The set

$$S^a = \{x \in G \mid axa^{-1} \in S\}$$

is called the *conjugate of S by a*. We note that $(S^a)^b = S^{ab}$ and that $S^e = S$.

Proposition. *If S is a subgroup of G, then S^a is also a subgroup of G.*

Proof. Suppose $x, y \in S^a$. Then $axa^{-1}, aya^{-1} \in S$, and consequently, since S is a subgroup, we have

$$(axa^{-1})(aya^{-1}) = axya^{-1} \in S.$$

Therefore, $xy \in S^a$. Similarly, $x \in S^a$ implies $axa^{-1} \in S$, from which it follows that

$$(axa^{-1})^{-1} = ax^{-1}a^{-1} \in S \quad \text{and} \quad x^{-1} \in S^a.$$

46. *Normal Subgroups.*   A subgroup H of a group G is *normal* if H is equal to each of its conjugates, that is, if $H^a = H$ for every element $a \in G$. Normal subgroups are also called *invariant* or *self-conjugate* subgroups. We write $H \lhd G$ to indicate that H is a normal subgroup of G.

It is easy to see and worthwhile to note that *every subgroup of an abelian group is normal*: if H is a subgroup of an abelian group G, then $x \in H^a$ if and only if $axa^{-1} = aa^{-1}x = x \in H$; thus, $H^a = H$ for all $a \in G$.

46α. Show that a subgroup H of a group G is normal if and only if every left coset of H is equal to some right coset of H.

46β. Show that a subgroup H of a group G is normal if and only if $ab \in H$ implies $a^{-1}b^{-1} \in H$ for any elements $a, b \in G$.

46γ. Show that a subgroup of index 2 is always normal.

46δ. Let H and N be subgroups of a group G and let N be normal. Show that $H \cap N$ is a normal subgroup of H. (This will be used in **69**.)

46ε. Let H and N be subgroups of a group G, N normal. Show that HN is a subgroup of G and that N is a normal subgroup of HN. (This will be used in **69**.)

46ζ. Let N and N' be normal subgroups of a group G. Show that NN' is a normal subgroup of G.

46η. Let $H, K,$ and N be subgroups of a group G, $K \lhd H$ and $N \lhd G$. Prove that $NK \lhd NH$.

46θ. Show that the center Z_G of a group G is always a normal subgroup. (See **35δ**.)

46ι. The *quaternion group* Q is a group of order 8 which may be presented as a group with two generators a and b subject to the relations $a^4 = e, b^2 = a^2$, and $aba = b$. To be explicit, the elements of Q are

$$e, a, a^2, a^3, b, ab, a^2b, a^3b$$

and the group product is given by Table 5.

Table 5

	e	a	a^2	a^3	b	ab	a^2b	a^3b
e	e	a	a^2	a^3	b	ab	a^2b	a^3b
a	a	a^2	a^3	e	ab	a^2b	a^3b	b
a^2	a^2	a^3	e	a	a^2b	a^3b	b	ab
a^3	a^3	e	a	a^2	a^3b	b	ab	a^2b
b	b	a^3b	a^2b	ab	a^2	a	e	a^3
ab	ab	b	a^3b	a^2b	a^3	a^2	a	e
a^2b	a^2b	ab	b	a^3b	e	a^3	a^2	a
a^3b	a^3b	a^2b	ab	b	a	e	a^3	a^2

(The entry in the *x*-row, *y*-column is the product *xy*.)

Prove that Q is *Hamiltonian*, that is, that Q is a non-abelian group of which every subgroup is normal.

46κ. Show that a non-abelian group of order 8 with a single element of order 2 is isomorphic to the quaternion group Q (**46ι**).

46λ. Show that the quaternion group Q is not isomorphic to the dihedral group D_4 (**26θ**).

46μ. Let H and K be normal subgroups of G such that $H \cap K = \{e\}$. Show that the group HK is isomorphic to $H \times K$, the direct product of H and K (**26ζ**).

46ν. Let H_1 and H_2 be subgroups of a group G and $N_1 \lhd H_1$, $N_2 \lhd H_2$. Show that

$$N_1(H_1 \cap N_2) \lhd N_1(H_1 \cap H_2)$$

and

$$(H_1 \cap N_2)(H_2 \cap N_1) \lhd (H_1 \cap H_2).$$

(This will be used in **70**.)

47.   Theorem.   *If H is a normal subgroup of G, then the left coset space G/H is a group with the product $(aH)(bH) = (ab)H$.*

Proof. Since the left cosets aH and bH are subsets of G, their product $(aH)(bH)$ is the set

$$\{x \in G \mid x = a'b' ; a' \in aH, b' \in bH\}.$$

This product depends only upon the cosets aH and bH and not upon the elements a and b. Therefore, it is well defined. Clearly, $(ab)H \subset (aH)(bH)$: given $x \in (ab)H$, $x = abh$ for some $h \in H$, and $x = a'b'$ where $a' = a \in aH$ and $b' = bh \in bH$. On the other hand, every element of $(aH)(bH)$ has the form $ahbh'$ where $h, h' \in H$. Since H is normal and $H^b = H$, we have $h = bh''b^{-1}$ for some $h'' \in H$. Therefore,

$$ahbh' = a(bh''b^{-1})bh' = (ab)(h''h') \in (ab)H.$$

Consequently, $(aH)(bH) = (ab)H$. In other words, we have shown that *when a subgroup is normal, the product of two of its left cosets is again a left coset*. This product is clearly associative, the left coset $eH = H$ serves as the identity, and $(aH)^{-1} = a^{-1}H$. It is now apparent that G/H is a group. G/H is called the *quotient group* of G by H.

Corollary. *If* $H \lhd G$ *and* G *is finite, then* $o(G/H) = o(G)/o(H)$.

This follows from **40** since $o(G/H) = [G:H]$.

47α. Describe the quotient group $\mathbf{Z}/m\mathbf{Z}$.

47β. $\mathbf{Z}$ is a normal subgroup of $\mathbf{R}$ (real numbers under addition). Show that the quotient group $\mathbf{R}/\mathbf{Z}$ is isomorphic to the circle group K (**31**).

47γ. Prove that every quotient group of a cyclic group is cyclic.

47δ. Let $H \lhd G$. Show that the order of aH as an element of the quotient group G/H divides the order of $a \in G$.

47ε. Prove that if G/Z_G is a cyclic group, then G is abelian. (Z_G denotes the center of G. See **35δ** and **46θ**.)

47ζ. Let $H \lhd G$. Show that G/H is abelian if and only if H contains every element of the form $aba^{-1}b^{-1}$, where $a, b \in G$.

47η. Let $H \lhd G$. Let K be a subgroup of G/H and set

$$\tilde{K} = \{g \in G \mid gH \in K\}.$$

Show that $\tilde{K}$ is a subgroup of G containing H and that $\tilde{K} \lhd G$ if and only if $K \lhd G/H$. Conclude that there is a one-to-one correspondence between subgroups (normal subgroups) of G/H and subgroups (normal subgroups) of G which contain H.

47θ. Let $G \times G'$ denote the direct product of the groups G and G' (**26ζ**). Show that $H \lhd G$ and $H' \lhd G'$ imply $(H \times H') \lhd (G \times G')$ and that the quotient group $(G \times G')/(H \times H')$ is isomorphic to the direct product $(G/H) \times (G'/H')$.

48. The *normalizer* of a subset S of a group G is the set

$$N(S) = \{a \in G \mid S^a = S\},$$

where S^a denotes the conjugate of S by a. (See **45**.)

Proposition. *The normalizer* $N(S)$ *of* S *in* G *is a subgroup of* G.

Proof.

(1) If $a, b \in N(S)$, then

$$S^{ab} = (S^a)^b = S^b = S,$$

and therefore $ab \in N(S)$.

(2) If $a \in N(S)$, then

$$S^{a^{-1}} = (S^a)^{a^{-1}} = S^{aa^{-1}} = S,$$

and therefore $a^{-1} \in N(S)$.

Proposition.   *If H is a subgroup of G, then $N(H)$ is the largest subgroup of G containing H as a normal subgroup.*

Proof.   Clearly $H \subset N(H)$, and therefore H is a subgroup of $N(H)$. To see that H is a normal subgroup of $N(H)$, we note that for $a \in N(H)$, the conjugate of H by a, *in the group $N(H)$*, is

$$\{x \in N(H) \,|\, axa^{-1} \in H\} = H^a \cap N(H) = H \cap N(H) = H.$$

Now suppose that H is normal in a subgroup N of G. In N, the conjugate of H by $a \in N$ is $N \cap H^a = H$, and therefore $H \subset H^a$. Similarly, $H \subset H^{a^{-1}}$, which implies that $H^a \subset (H^a)^{a^{-1}} = H$. Thus, for every $a \in N$ we have $H^a = H$ and $a \in N(H)$, or in other words, we have shown that $N \subset N(H)$. This is what is meant by saying that $N(H)$ is the largest subgroup of G in which H is normal.

48α.   Using the description of the dihedral group D_6 given in **35ζ**, compute the normalizer of the sets $\{\alpha\}$, $\{\beta\}$, and $\langle \alpha \rangle = \{e, \alpha, \alpha^2, \alpha^3, \alpha^4, \alpha^5\}$.

48β.   Determine the normalizer in the symmetric group S_n of the subgroup H of all permutations leaving n fixed. Determine all the conjugates of H in S_n.

48γ.   Show that $N(S^a) = N(S)^a$ for any subset S of a group G.

49.   Theorem.   *The number of distinct subsets of a group G which are conjugates of a given subset S is $[G : N(S)]$, the index in G of the normalizer of S.*

Proof. $S^a = S^b$ if and only if $S^{ab^{-1}} = S$. In other words, $S^a = S^b$ if and only if $ab^{-1} \in N(S)$, which is equivalent to $a^{-1} \equiv b^{-1}$ mod $N(S)$. This sets up a one-to-one correspondence from the class $\mathscr{C}(S)$ of subsets of G conjugate to S, to the left coset space $G/N(S)$ given by $S^a \to a^{-1}N(S)$. Since $G/N(S)$ has $[G : N(S)]$ elements, the theorem follows.

49α.   Let S be a subset of a group G, which has exactly two conjugates. Show that G has a proper nontrivial normal subgroup.

49β.   Let H be a proper subgroup of *a finite group G*. Show there is at least one element of G not contained in H or in any of its conjugates.

50. The *center* of a group G is the set

$$Z_G = \{x \in G \mid xa = ax, \text{ for all } a \in G\}.$$

Proposition. *The center of a group is a normal subgroup.*

Proof. If $x, y \in Z_G$, then for any $a \in G$ we have $xya = xay = axy$, so that $xy \in Z_G$. If $x \in Z_G$, then for any $a \in G$, $xa^{-1} = a^{-1}x$, which implies that

$$x^{-1}a = (a^{-1}x)^{-1} = (xa^{-1})^{-1} = ax^{-1},$$

so that $x^{-1} \in Z_G$. Thus, Z_G is a subgroup of G. For any $a \in G$ we have $x \in Z_G^a$ if and only if $axa^{-1} = xaa^{-1} = x \in Z_G$. Therefore, $Z_G^a = Z_G$ for all $a \in G$, and Z_G is a normal subgroup of G.

Proposition. *For any group G,* $Z_G = \{x \in G \mid N(x) = G\}$.

Proof. $N(x)$ denotes the normalizer of the set $\{x\}$. If $N(x) = G$, then $\{x\}^a = \{x\}$ for each $a \in G$, or what is the same thing, $axa^{-1} = x$ and $ax = xa$ for all $a \in G$, which means $x \in Z_G$. Reversing all these implications shows that $N(x) = G$ implies $x \in Z_G$.

50α. Show that the center of the symmetric group S_n is trivial for $n > 2$.

50β. Compute the center of the dihedral group D_n. (See **26θ** and **35η**.)

50γ. Compute the center of the quaternion group Q (**46ι**).

51. *The Conjugacy Class Equation of a Group.* An element x of a group G is *conjugate* to an element y of G if $x = aya^{-1}$ for some element a of G—or what is the same thing, if the set $\{y\}$ is a conjugate of the set $\{x\}$ as defined in **45**. It is not difficult to verify that conjugacy is an equivalence relation and divides G into disjoint *conjugacy classes*. The number of elements in C_x, the conjugacy class of x, is $[G: N(x)]$, the index in G of $N(x)$, the normalizer of $\{x\}$. (This follows from **49**, since the number of elements in C_x is the number of conjugates of the set $\{x\}$.) Clearly, $x \in Z_G$, the center of G, if and only if $N(x) = G$, $[G: N(x)] = 1$, and $C_x = \{x\}$.

Suppose now that G is a finite group, and that consequently there are a finite number of conjugacy classes. By the above remarks it is clear that either $x \in Z_G$ or $C_x \subset G - Z_G$. Let $x_1, x_2, \ldots, x_m \in G$ be elements obtained by choosing one element from each conjugacy class contained in $G - Z_G$. For any $x \in G$, either $x \in Z_G$ or x is conjugate to one x_i. Counting up the elements of G, we conclude that

$$o(G) = o(Z_G) + \sum_{i=1}^{m} [G: N(x_i)].$$

This is called the *conjugacy class equation* of G.

51α. Divide the elements of the quaternion group Q into conjugacy classes and verify the conjugacy class equation.

51β. Prove that a group of order p^2 is abelian (p a prime).

52. Prime Power Groups. We illustrate the results of the last few articles with a brief study of groups whose order is a power of a prime number. Let G be a group of order p^n, where p is a prime number and $n > 1$.

Lemma 1. *G has a nontrivial center.*

Proof. In the conjugacy class equation of G

$$o(G) = o(Z_G) + \sum_{i=1}^{m} [G: N(x_i)],$$

we must have $[G: N(x_i)] > 1$ so that $p \mid [G: N(x_i)]$ for each $i = 1, 2, \ldots, m$. Since $p \mid o(G)$, it follows that $p \mid o(Z_G)$ and therefore $Z_G \neq \{e\}$.

Lemma 2. *G has a proper nontrivial normal subgroup.*

Proof. Z_G, the center of G, is a normal subgroup of G and is nontrivial by lemma 1. If Z_G is proper, we are finished. If $Z_G = G$, then G is abelian and any element $a \in G$, $a \neq e$, generates a nontrivial normal subgroup $\langle a \rangle$. If $o(a) = o(\langle a \rangle) < o(G)$, then we are finished. On the other hand, $o(a) = o(G) = p^n$ implies $o(a^p) = p^{n-1}$, and a^p generates a proper nontrivial normal subgroup.

Theorem. *If G is a group of order p^n, where p is a prime number and $n \geq 1$, then G has a sequence of subgroups $\{e\} = G_0 \subset G_1 \subset \cdots \subset G_n = G$ such that $o(G_k) = p^k$, G_k is normal in G_{k+1}, and the quotient group G_{k+1}/G_k is cyclic of order p.*

Proof. The proof is by induction on the integer n. For $n = 1$ it is trivial because $\{e\} = G_0 \subset G_1 = G$, and a group of order p, p prime, is cyclic (**43**). Suppose the theorem is true for all groups of order p^k for $k < n$, and let G be a group of order p^n. By lemma 2, G has a proper nontrivial normal subgroup. Among all such subgroups, finite in number, let H be one of maximal order, say $o(H) = p^t$, $t < n$. We want to show $t = n - 1$.

Suppose $t < n - 1$; then G/H is a group of order $p^{n-t} \geq p^2$. By lemma 2, G/H has a proper, nontrivial, normal subgroup N. Let H' denote the subset $\{g \in G \mid gH \in N\}$ of G. H' is a normal subgroup of G:

(1) If $g, g' \in H'$, then $gH, g'H \in N$, and since N is a subgroup of G/H, $(gH)(g'H) = gg'H \in N$ and $gg' \in H'$.

(2) If $g \in H$, then $gH \in N$, and since N is a subgroup, $(gH)^{-1} = g^{-1}H \in N$ so that $g^{-1} \in H'$.

(3) $(H')^a = \{g \in G \mid aga^{-1} \in H'\} = \{g \in G \mid aga^{-1}H \in N\}$
$$= \{g \in G \mid (aH)(gH)(aH)^{-1} \in N\}$$
$$= \{g \in G \mid gH \in N^{aH}\}$$
$$= \{g \in G \mid gH \in N\}$$
$$= H'.$$

Clearly, $H'/H = N$ and therefore

$$o(H')/o(H) = [H' : H] = o(N) \geq p,$$

or in other words, $o(H') \geq po(H)$ contradicting maximality of H. Consequently, the assumption that $t < n - 1$ is incorrect, and $t = n - 1$.

To finish the proof we set $G_{n-1} = H$ and apply the inductive assumption to the group G_{n-1} of order p^{n-1} to obtain the groups $G_0 \subset G_1 \subset \cdots \subset G_{n-1}$. Since $G/H = G_n/G_{n-1}$ is a group of order p and p is prime, it is automatically cyclic (**43**).

The sequence

$$\{e\} = G_0 \subset G_1 \subset \cdots \subset G_{n-1} \subset G_n = G$$

is called a *composition series* for G. (See **73**.)

he Sylow Theorems

Some information about the structure of a finite group can be obtained from its order alone. The most important results in this direction are the three theorems of Sylow which are proved in the next few articles. First we give a basic theorem on transformation groups which will simplify the proofs.

53. Transformation Groups. A group G *acts* on a set X (as a *group of transformations*) if to each pair $(g, x) \in G \times X$ there is associated an element $g * x \in X$ in such a way that

(1) $g * (h * x) = (gh) * x$ for all $g, h \in G$ and all $x \in X$;

(2) $e * x = x$ for all $x \in X$. (e is the identity element of G.)

We note that each $g \in G$ determines a one-to-one correspondence $g: X \to X$, given by $g(x) = g * x$, whose inverse is $g^{-1}: X \to X$. (These one-to-one correspondences are sometimes called *transformations* of X.)

As examples of transformation groups we note that every group G acts on itself by the rule $g * h = gh$ for all $g, h \in G$, and more generally, if H is a subgroup of G, then G acts on the left coset space $X = G/H$ by the rule $g * (g'H) = (gg')H$.

53α. Explain how the symmetric group S_n acts as a group of transformations on any set X with n elements.

53β. Let G be a group of transformations of a set X. Define G_0 by

$$G_0 = \{g \in G \mid g * x = x, \text{ all } x \in X\}.$$

Show that G_0 is a normal subgroup of G.

53γ. A group of transformations G of a set X acts *effectively* on X if $g * x = x$ for all $x \in X$ implies $g = e$. In other words, G is an effective transformation group of X whenever $G_0 = \{e\}$. (See **53β**.) Show that if G is any transformation group of X, effective or not, the quotient group G/G_0 acts effectively on X.

53δ. Show that a group G acts effectively on the left coset space G/H if and only if $\bigcap_{a \in G} H^a = \{e\}$.

53ε. Show that the set $\mathcal{A}(X)$ of one-to-one correspondences of a set X forms the largest effective transformation group of X.

54. Orbits and Stabilizers. Let G be a group acting on the set X. We define an equivalence relation $\sim$ on X by setting $x \sim y$ if and only if $y = g * x$ for some $g \in G$. An equivalence class under $\sim$ is called an *orbit*. The orbit of $x \in X$ is simply the set

$$G * x = \{y \in X \mid y = g * x, \text{ for some } g \in G\}.$$

The quotient set $X/\sim$ is called the *set of orbits of X under the action of G*. For each $x \in X$, the set

$$G_x = \{g \in G \mid g * x = x\}$$

is a subgroup of G called the *stabilizer* of x.

Theorem. *If G is a group acting on a finite set X, then the number of elements in the orbit of $x \in X$ is the index in G of the stabilizer G_x.*

Proof. The mapping $\phi: G \to G * x$ given by $\phi(g) = g * x$ is clearly onto. Furthermore, $\phi(g) = \phi(h)$ if and only if $g^{-1}h \in G_x$, or what is the same thing, $g \equiv h \bmod G_x$. This implies there is a well-defined, one-to-one correspondence $G/G_x \to G * x$ given by $gG_x \mapsto g * x$. It follows that $G * x$ has the same number of elements as G/G_x, namely $[G: G_x]$.

One illustration of this theorem is as follows. Let 2^G denote the power set of a finite group G, that is, the set whose elements are the subsets of G. (It is not hard to see that 2^G has $2^{o(G)}$ elements.) Let G act on 2^G by the rule

$$y * S = S^{g^{-1}} = \{x \in G \mid g^{-1}xg \in S\}.$$

Then the orbit of a set $S \subset G$ is just $\mathscr{C}(S)$, the collection of conjugates of S, and the stabilizer of S is just its normalizer $N(S)$. Thus, the theorem above implies **49** as a special case—the number of conjugates of S is $[G: N(S)]$.

54α. Let H be a subgroup of a group G. Then H acts on G by the rule $h * x = hx$ for $h \in H$, $x \in G$. What is the orbit of $x \in G$ under this action? What is the stabilizer H_x?

54β. Let G be a group acting on a set X. Show that two elements of X which belong to the same orbit have conjugate subgroups of G as stabilizers.

54γ. Let H and K be subgroups of a group G. K acts on the left coset space G/H by $k * (gH) = (kg)H$. What is the stabilizer of gH? What is the number of elements in the orbit of gH?

55. Cauchy's Theorem. *If p is a prime dividing the order of a finite group G, then G has an element of order p.*

Proof. Let n be the order of G. Let

$$X = \{(a_1, a_2, \ldots, a_p) \in G^p \mid a_1 a_2 \cdots a_p = e\}.$$

(G^p denotes the cartesian product $G \times G \times \cdots \times G$ with p factors.) Since the first $p - 1$ coordinates of an element of X may be chosen arbitrarily from G, thereby determining the last one, it is clear that X has n^{p-1} elements.

Let C be a cyclic group of order p generated by the element c. Let C act on the set X by the rule

$$c * (a_1, a_2, \ldots, a_p) = (a_2, a_3, \ldots, a_p, a_1).$$

By **54** the number of elements in any orbit divides the order of C, which is p. Thus, an orbit has either p elements or one element. Let r be the number of orbits with one element, and let s be the number of orbits with p elements. Then $r + sp = n^{p-1}$, the number of elements of X.

By hypothesis $p \mid n$, and consequently, $r + sp = n^{p-1}$ implies that $p \mid r$. We know that $r \neq 0$ because the orbit of $(e, e, \ldots, e) \in X$ has only one element. Therefore, there are at least p orbits with a single element. Each such element has the form $(a, a, \ldots, a) \in X$ so that $a^p = e$. Thus, G contains at least p elements solving the equation $x^p = e$. Clearly then, G contains an element $a \neq e$ such that $a^p = e$, and a must have order p.

Remarks.   This beautiful proof of Cauchy's theorem is due to James H. McKay ("Another proof of Cauchy's group theorem," *American Mathematical Monthly*, vol. 66 (1959), p. 119).

Augustin Louis, Baron Cauchy (1789–1857) was a French mathematician whose prodigious contributions are important to all branches of mathematics. Among his contributions to algebra, in addition to the theorem above, is a theorem concerning the number of distinct values assumed by a function of several variables when the variables are permuted. This theorem lies at the crux of Abel's argument that algebraic equations of the fifth degree are not generally solvable by radicals. (See **85α** and **149**.) Cauchy is perhaps the one mathematician entitled to be called the founder of group theory.

56.   First Sylow Theorem.   *If p is prime and p^n divides the order of a finite group G, then G has a subgroup of order p^n.*

Proof.   Suppose $o(G) = p^n m$. The number of subsets of G which have precisely p^n elements is the binomial coefficient

$$N = \binom{p^n m}{p^n} = \frac{(p^n m)(p^n m - 1) \cdots (p^n m - i) \cdots (p^n m - p^n + 1)}{(p^n)(p^n - 1) \cdots (p^n - i) \cdots (1)}.$$

(This is the number of ways that a set of p^n elements can be chosen from a set of $p^n m$ elements.) For $0 < i < p^n$ the highest power of p dividing $p^n m - i$ is the same as that dividing $p^n - i$. (Why?) Thus, all the factors of p in the numerator and the denominator of N cancel out, *except those of m*. Consequently, m and N have the same number of factors of p. Let p^r be the highest power of p dividing m and N.

G acts on the set $\mathscr{S}$ of all subsets of G with p^n elements by the rule

$$g * S = gS = \{x \in G \mid x = gs; s \in S\}$$

for any $S \in \mathscr{S}$. If every orbit under this action were divisible by p^{r+1}, we would have $p^{r+1} \mid N$. (Why?) Therefore, there is at least one orbit, say

$\{S_1, S_2, \ldots, S_k\}$, for which $p^{r+1} \nmid k$. Let H denote the stabilizer of S_1. We know that $k = [G : H] = o(G)/o(H)$ by **54**. Since $p^r \mid m$, we have $p^{n+r} \mid p^n m$. However, $p^n m = o(G) = ko(H)$, and therefore, we have $p^{n+r} \mid ko(H)$. Since $p^{r+1} \nmid k$, we must conclude that $p^n \mid o(H)$ and that $p^n \leq o(H)$.

On the other hand, because H stabilizes S_1 we have $Hg \subset S_1$ for any $g \in S_1$. Thus,

$$o(H) = \ast(Hg) \leq \ast(S_1) = p^n.$$

Combined with $p^n \leq o(H)$, this implies $o(H) = p^n$ and the proof is complete.

Remarks. This elegant argument, which is a great improvement over the older method of double cosets, comes from a paper of Helmut Wielandt, "Ein Beweis für die Existenz der Sylowgruppen," *Archiv der Matematik*, vol. 10 (1959) pp. 401–402. The original theorem of the Norwegian mathematician Ludwig Sylow (1832–1918) stated only the existence of a subgroup of order p^n where p^n is the highest power of p dividing the order of the group. Such a subgroup is called a *p-Sylow subgroup*. Of course the more general statement above is easily deduced from Sylow's theorem and the structure of prime power groups (**52**). Finally, we observe that Cauchy's theorem. (**55**) is a special case of the first Sylow theorem, since the existence of a subgroup of order p implies the existence of an element of order p.

57. Second Sylow Theorem. *All the p-Sylow subgroups of a finite group are conjugate.*

Proof. Let G be a finite group of order $p^n m$, where $p \nmid m$ and $n > 0$. Let H be a p-Sylow subgroup of G. Of course $o(H) = p^n$ and $[G : H] = m$. Let $S_1, S_2, \ldots, S_m$ denote the left cosets of G mod H. G acts on G/H by the rule $g * S_i = gS_i$. Let H_i denote the stabilizer of S_i.

All the groups H_i are conjugates of H. To see this we note that by **54**, $o(H_i) = p^n$, while $gHg^{-1} \subset H_i$ if $S_i = gH$. Since $o(gHg^{-1}) = o(H) = o(H_i)$, we have $gHg^{-1} = H_i$.

Let H' be a second p-Sylow subgroup of G. Then H' also acts on G/H by the same rule as G. Since $p \nmid m$, there is at least one orbit (under H') with a number of elements not divisible by p. We may suppose that $S_1, S_2, \ldots, S_r$ are the elements of an orbit where $p \nmid r$. Let $K = H' \cap H_1$. Then K is the stabilizer of S_1 under the action of H'. Therefore, $[H' : K] = r$. However, $o(H') = p^n$ and $p \nmid r$, from which it follows that $r = 1$ and $K = H'$. Therefore, $o(K) = o(H') = o(H_1) = p^n$, and $H' = K = H_1$. Thus, H' and H are conjugate.

58. Third Sylow Theorem. *The number of p-Sylow subgroups of a finite group is a divisor of their common index and is congruent to 1 modulo p.*

Proof.   Let G be a group of order $p^n m$, where $n > 0$ and $p \nmid m$. Suppose r is the number of p-Sylow subgroups of G. Then we want to show that $r \mid m$ and that $r \equiv 1 \bmod p$.

As before, let H be any one of the p-Sylow subgroups of G. Of course $o(H) = p^n$ and $[G : H] = m$. We shall denote the elements of the left coset space G/H by $S_1, S_2, \ldots, S_m$. G acts on G/H by the rule $g * S = gS$ for $S \in G/H$. There is only one orbit under this action, namely the whole of G/H. (Why?) Therefore the stabilizer of each S_i is a subgroup in G of index m and order p^n. In other words, each coset S_i has a p-Sylow subgroup as stabilizer.

On the other hand, as we shall soon see, each p-Sylow subgroup is the stabilizer of one or more of the cosets $S_1, S_2, \ldots, S_m$. Clearly, H is the stabilizer of the coset H, which must occur among the S_i's. Let $S_1, S_2, \ldots, S_k$ be the elements of G/H whose stabilizer is H. By the second Sylow theorem any other p-Sylow subgroup of G is a conjugate gHg^{-1} of H. It is easy to see that gHg^{-1} stabilizes the cosets $gS_1, gS_2, \ldots, gS_k$. Consequently, we see that each one of the r distinct p-Sylow subgroups of G is the stabilizer of exactly k elements of G/H. Hence, we conclude that $m = kr$ and that $r \mid m$.

Now we restrict our attention to just the action of H on G/H. Unless $H = G$ and $r = 1$, there is more than one orbit. (Why?) Applying the orbits and stabilizers theorem of **54** to this restricted action and using the fact that $o(H) = p^n$, we can distinguish two cases:

(1) the orbit of S_i contains p^t elements for some t, $0 < t < n$;
(2) the orbit of S_i contains only the element S_i.

Clearly, the second case occurs if and only if S_i is one of the cosets $S_1, S_2, \ldots, S_k$ whose stabilizer is H. Thus, counting the elements of G/H, we conclude that $m = k + up$ or that $m \equiv k \bmod p$.

The previous conclusion that $m \equiv kr$ along with $m \equiv k \bmod p$ yields $kr \equiv k \bmod p$, from which it follows that $r \equiv 1 \bmod p$, since $k \not\equiv 0 \bmod p$. The proof is accomplished.

59.   As a simple example of the direct application of the Sylow theorems, we note that *a group of order* 100 *has a normal subgroup of order* 25. The first theorem guarantees a 5-Sylow subgroup of order 25. The number of such subgroups is congruent to 1 modulo 5 and divides 4 by the third theorem. Thus, there is only one such subgroup of order 25. Since this subgroup equals each of its conjugates, it must be normal.

59α.   Let G be a group of order pq, where p and q are prime and $p < q$. Show that G is cyclic when $q \not\equiv 1 \bmod p$. What can be said of G when $q \equiv 1 \bmod p$?

59β.   Show that a group of order $2p$, where p is prime, is either cyclic or isomorphic to the dihedral group D_p.

59γ. Suppose that G is a finite group with a normal subgroup H of order p^k, where p is prime and $k > 0$. Show that H is contained in every p-Sylow subgroup of G.

59δ. Let H be a normal subgroup of finite index in a group G. Show that if $p^k \mid [G : H]$, where p is prime, then G contains a subgroup K such that $[K : H] = p^k$.

59ε. A group is *simple* if it has only itself and the trivial group as normal subgroups. (For example, groups of prime order are obviously simple.) Prove that a group of order 30 cannot be simple.

59ζ. Show that a group of order $p^2 q$, where p and q are primes, is not simple (**59ε**).

59η. Show that a simple group of order less than 60 is of prime order. (We shall see later that there does exist a simple group of order 60, the alternating group A_5.)

59θ. Let G be a finite group with just one p-Sylow subgroup for each prime p dividing $o(G)$. Show that G is isomorphic to the direct product of all its Sylow subgroups. (See **26ζ** and **26ν**.)

59ι. Let p be a prime such that p^k divides the order of the finite group G. Prove that the number of subgroups of order p^k in G is congruent to 1 modulo p. (This is a theorem of Georg Frobenius (1849–1917), and it is only fair to warn that the solution is somewhat lengthy.)

oup Homomorphism
d Isomorphism

---

60. A *homomorphism* of groups is a mapping from the set of elements of one group to the set of elements of another which preserves multiplication. In other words, a mapping $\phi : G \to G'$ is a group homomorphism if G and G' are groups and if for all $x, y \in G$, $\phi(xy) = (\phi x)(\phi y)$.

A group homomorphism $\phi : G \to G$ is called an *endomorphism* (of the group G). The identity mapping $1_G : G \to G$ of any group G is clearly an endomorphism.

The composition of homomorphisms is again a homomorphism. Indeed, if $\phi : G \to G'$ and $\psi : G' \to G''$ are group homomorphisms, then for all $x, y \in G$ we have

$$(\psi\phi)(xy) = \psi(\phi(xy)) = \psi((\phi x)(\phi y)) = \psi(\phi x)\psi(\phi y) = ((\psi\phi)x)((\psi\phi)y),$$

and therefore $\psi\phi: G \to G''$ is a homomorphism.

60α. Show that a group homomorphism preserves identity elements and inverses. That is, show that $\phi e = e'$ and that $\phi(g^{-1}) = (\phi g)^{-1}$ for any homomorphism $\phi: G \to G'$ of groups with identity elements e and e'.

60β. Let H be a normal subgroup of a group G. Show that the mapping $\phi: G \to G/H$ given by $\phi g = gH$ is a group homomorphism.

60γ. Show that a group G is abelian if and only if the mapping $\phi: G \to G$ given by $\phi g = g^{-1}$ is an endomorphism of G.

60δ. Show that a group G is abelian if and only if the mapping $\phi: G \to G$ given by $\phi g = g^2$ is an endomorphism of G.

60ε. Show that a group G is abelian if and only if the mapping $\phi: G \times G \to G$ given by $\phi(a, b) = ab$ is a group homomorphism.

60ζ. Determine the number of distinct homomorphisms $\phi: \mathbf{Z}_m \to \mathbf{Z}_n$ in terms of m and n.

61.   Proposition.   *A mapping $\phi: \mathbf{Z} \to \mathbf{Z}$ is an endomorphism of the additive group of integers if and only if there is an integer k such that $\phi(n) = kn$ for all $n \in \mathbf{Z}$.*

Proof.   Let $k \in \mathbf{Z}$ and $\phi(n) = kn$ for all $n \in \mathbf{Z}$. Then for all $n, m \in \mathbf{Z}$ we have

$$\phi(n + m) = k(n + m) = kn + km = \phi(n) + \phi(m),$$

so that ϕ is a group homomorphism from $\mathbf{Z}$ to $\mathbf{Z}$. Suppose on the other hand that $\phi: \mathbf{Z} \to \mathbf{Z}$ is an endomorphism. Let $k = \phi(1)$. Since $n = 1 + \cdots (n) \cdots + 1$ for a positive integer n, we have

$$\phi(n) = \phi(1) + \cdots (n) \cdots + \phi(1) = k + \cdots (n) \cdots + k = kn.$$

Also,

$$\phi(1) = \phi(1 + 0) = \phi(1) + \phi(0)$$

implies $\phi(0) = 0 = k \cdot 0$, from which it follows that

$$\phi(-1) = -k \quad \text{and} \quad \phi(-n) = -kn$$

for any positive n. Thus, for all $n \in \mathbf{Z}$ we have $\phi(n) = kn$.

62. An *isomorphism* $\phi: G \to G'$ is a one-to-one correspondence which preserves group multiplication. An isomorphism is therefore a very special kind of homomorphism. If $\phi: G \to G'$ is an isomorphism, it is easy to show (as the reader should) that the inverse mapping $\phi^{-1}: G' \to G$ preserves multiplication and is again an isomorphism.

A group G is *isomorphic* to a group G' if there exists an isomorphism $\phi: G \to G'$. This is denoted by writing $G \approx G'$ or $\phi: G \approx G'$. Isomorphism is an equivalence relation among groups:

(1) $G \approx G$ for any group G;
(2) $G \approx G'$ implies $G' \approx G$; and
(3) $G \approx G'$, $G' \approx G''$ imply $G \approx G''$.

Isomorphic groups have the same structure and the same group-theoretic properties. In a sense, group theory is the study of those properties of groups which are preserved under isomorphism.

62α. Let $\mathbf{R}$ denote the group of all real numbers under addition and let $\mathbf{R}^+$ denote the group of all positive real numbers under multiplication. Show that the mapping $\phi: \mathbf{R} \to \mathbf{R}^+$ given by $\phi x = e^x$ is an isomorphism. What is the inverse of ϕ?

62β. Let $\phi: \mathbf{Z}_{16} \to \mathbf{Z}'_{17}$ be given by $\phi[k]_{16} = [3^k]_{17}$. Show that ϕ is an isomorphism. (See **33** and **34**.)

62γ. Show that a group of order 8 is isomorphic to $\mathbf{Z}_8$ (**33**), D_4 (**26θ**), Q (**46ι**), $\mathbf{Z}_4 \times \mathbf{Z}_2$, or $\mathbf{Z}_2 \times \mathbf{Z}_2 \times \mathbf{Z}_2$.

62δ. Show that a group of order p^2, where p is prime, is isomorphic to $\mathbf{Z}_{p^2}$ or $\mathbf{Z}_p \times \mathbf{Z}_p$.

62ε. Let G denote the group of real numbers between -1 and $+1$ under the operation $x \cdot y = (x + y)/(1 + xy)$. Show that G is isomorphic to the group of real numbers $\mathbf{R}$ under addition.

62ζ. Show that every finite abelian group is isomorphic to a direct product of cyclic groups.

63. Proposition. *A group G of order n is cyclic if and only if $G \approx \mathbf{Z}_n$.*

Proof. If G is a cyclic group of order n (**43**), then $G = \{e = a^0, a^1, \ldots, a^{n-1}\}$. Recall that $\mathbf{Z}_n$, the additive group of integers modulo n, may be represented as the set $\{[0]_n, [1]_n, \ldots, [n-1]_n\}$ (**33**). An isomorphism $\phi: G \to \mathbf{Z}_n$ is given by $\phi(a^k) = [k]_n$, since ϕ is clearly a one-to-one correspondence and

$$\phi(a^j a^k) = \phi(a^{j+k}) = [j + k]_n = [j]_n + [k]_n = \phi(a^j) + \phi(a^k).$$

(This isomorphism is not unique—it depends on the choice of a generator for G.)

On the other hand, suppose G is a group and $\phi: \mathbf{Z}_n \to G$ an isomorphism. Let $a = \phi[1]_n$. If $g \in G$, then $g = \phi[k]_n$ for some $[k]_n \in \mathbf{Z}_n$, and therefore

$$g = \phi[k]_n = \phi([1]_n + \cdots (k) \cdots + [1]_n) = (\phi[1]_n)^k = a^k.$$

This shows that every element of G is a power of a and hence that G is cyclic.

64.   An *automorphism* of a group is an isomorphism of the group with itself. The identity $1_G: G \to G$ of a group is an automorphism of G, and in general there are many others.

Let a be an element of a group G, and let $\alpha_a: G \to G$ be the mapping given by $\alpha_a(g) = a^{-1}ga$ for every $g \in G$. Then α_a is a homomorphism since

$$\alpha_a(gg') = a^{-1}gg'a = (a^{-1}ga)(a^{-1}g'a) = (\alpha_a(g))(\alpha_a(g')).$$

To prove that α_a is an isomorphism, and hence an automorphism, it is sufficient to observe that $\alpha_{a^{-1}}$ is an inverse for α_a. Thus, to each element $a \in G$ there is assigned an automorphism α_a. Such automorphisms are called *inner automorphisms*. All other automorphisms (if there are any) are called *outer automorphisms*.

If S is a subset of a group G and $a \in G$, then the conjugate of S by a is the set

$$S^a = \{x \in G \mid axa^{-1} \in S\}$$

(45). It follows that $S^a = \alpha_a(S)$, since $x \in S^a$ means $axa^{-1} = y \in S$, or $x = a^{-1}ya \in \alpha_a(S)$ and vice versa. The point is that we may use this to give a variation of the definition of normal subgroup: *a subgroup H of a group G is normal if $\alpha_a(H) = H$ for all $a \in G$.* (This is the source of the term *invariant subgroup* for normal subgroups.)

64α.   Show that the set $\mathscr{A}(G)$ of all automorphisms of a group G is a group under composition and that the set $\mathscr{I}(G)$ of inner automorphisms of a group G is a normal subgroup of $\mathscr{A}(G)$. (The quotient group $\mathscr{A}(G)/\mathscr{I}(G)$ is called the *group of outer automorphisms* of G, which is a misnomer because the elements of $\mathscr{A}(G)/\mathscr{I}(G)$ are not outer automorphisms nor do the outer automorphisms themselves form a group.)

64β.   Show that the quotient of a group by its center is isomorphic to its group of inner automorphisms, that is, $G/Z_G \approx \mathscr{I}(G)$.

64γ.   Show that for any finite group G, $o(G) > 2$ implies $o(\mathscr{A}(G)) > 1$.

64δ.   Determine the number of distinct automorphisms of the groups $\mathbf{Z}_n$ and D_n.

64ε. A subgroup H of a group G is *characteristic* if $\phi(H) = H$ for all $\phi \in \mathscr{A}(G)$. Show that K a characteristic subgroup of H and H a characteristic subgroup of G imply that K is a characteristic subgroup of G. Show that K characteristic in H and H normal in G imply K normal in G.

64ζ. Determine the group $\mathscr{A}(\mathbf{Z}_2 \times \mathbf{Z}_2)$.

65. If $\phi: G \to G'$ is a group homomorphism, then the *kernel* of ϕ is the set

$$\text{Ker } \phi = \{x \in G \mid \phi x = e' \in G'\},$$

where e' is the identity element of G'. In other words the kernel of ϕ is the set of elements of G which are mapped into the identity element e' of G'. Since $x = xe$ for any $x \in G$, we have $\phi x = (\phi x)(\phi e)$, which implies $\phi e = e'$. (Why?) Therefore, $e \in \text{Ker } \phi$ and $\text{Ker } \phi$ is not empty.

Proposition. *For any homomorphism of groups $\phi: G \to G'$, $\text{Ker } \phi$ is a normal subgroup of G.*

Proof. First, we note that for all $x \in G$,

$$\phi(x^{-1}) = \phi(x)^{-1}$$

because $xx^{-1} = e$ implies

$$(\phi(x))(\phi(x^{-1})) = \phi(e) = e'.$$

From this it follows that if $x \in \text{Ker } \phi$, then

$$\phi(x^{-1}) = (\phi(x))^{-1} = e'^{-1} = e',$$

so that $x^{-1} \in \text{Ker } \phi$. If $x, y \in \text{Ker } \phi$, then

$$\phi(xy) = (\phi x)(\phi y) = e'e' = e',$$

so that $xy \in \text{Ker } \phi$. Thus, $\text{Ker } \phi$ is a subgroup of G.

Let $x \in (\text{Ker } \phi)^a$, the conjugate of $\text{Ker } \phi$ by $a \in G$. Then, $axa^{-1} \in \text{Ker } \phi$ and $(\phi a)(\phi x)(\phi a^{-1}) = e'$ which implies $\phi x = e'$ and $x \in \text{Ker } \phi$, or in other words, $(\text{Ker } \phi)^a \subset \text{Ker } \phi$. On the other hand, if $x \in \text{Ker } \phi$, then $axa^{-1} \in \text{Ker } \phi$ and $x \in (\text{Ker } \phi)^a$, or $\text{Ker } \phi \subset (\text{Ker } \phi)^a$. Thus, $\text{Ker } \phi$ is normal.

Proposition. *A homomorphism $\phi: G \to G'$ is one to one if and only if $\text{Ker } \phi = \{e\}$.*

Proof. If ϕ is one to one, then $\text{Ker } \phi = \phi^{-1}\{e'\}$ can contain but one element of G, and since $\phi(e) = e'$, it must be that $\text{Ker } \phi = \{e\}$. Conversely, if $\text{Ker } \phi = \{e\}$ and $\phi(a) = \phi(b)$ for $a, b \in G$, then

$$\phi(ab^{-1}) = \phi(a)\phi(b)^{-1} = e',$$

so that $ab^{-1} \in \mathrm{Ker}\ \phi$, $ab^{-1} = e$, and $a = b$. This shows that ϕ is one to one. (Why?)

A homomorphism which is one to one is called a *monomorphism*.

66. The *image* **(11)** of a homomorphism $\phi: G \to G'$ is the set

$$\mathrm{Im}\ \phi = \{x \in G' \mid x = \phi(y), y \in G\}.$$

Proposition. *For any homomorphism of groups* $\phi: G \to G'$, $\mathrm{Im}\ \phi$ *is a subgroup of* G'.

The proof is a simple exercise for the reader.

Proposition. *If* $\phi: G \to G'$ *is a group homomorphism, then*

$$G/\mathrm{Ker}\ \phi \approx \mathrm{Im}\ \phi.$$

Proof. We recall that the quotient group $G/\mathrm{Ker}\ \phi$ is the set of left cosets $a(\mathrm{Ker}\ \phi)$ with the product $a(\mathrm{Ker}\ \phi)b(\mathrm{Ker}\ \phi) = (ab)(\mathrm{Ker}\ \phi)$. (See **47.**) Let $\beta: G/\mathrm{Ker}\ \phi \to \mathrm{Im}\ \phi$ be the mapping given by $\beta(a(\mathrm{Ker}\ \phi)) = \phi(a)$. We must first check that β is *well defined*: a given element $a(\mathrm{Ker}\ \phi)$ of $G/\mathrm{Ker}\ \phi$ may be written in many ways, but the definition of β is given in terms of a specific expression. Suppose that $a(\mathrm{Ker}\ \phi) = b(\mathrm{Ker}\ \phi)$. Then $b \in a(\mathrm{Ker}\ \phi)$ and $b = ac$, where $c \in \mathrm{Ker}\ \phi$. It then follows that

$$\phi(b) = \phi(ac) = \phi(a)\phi(c) = \phi(a)e' = \phi(a).$$

Thus, the value of β on a coset does not depend on the specific representation of the coset—it is the same for all representations.

It is clear from the definition of β that it is a homomorphism, and it only remains to see that β is a one-to-one correspondence. If $x \in \mathrm{Im}\ \phi$, then $x = \phi(y)$ for some $y \in G$, and $x = \beta(y(\mathrm{Ker}\ \phi))$. Therefore, β is onto. On the other hand, it is immediate from the formula for β that $\mathrm{Ker}\ \beta$ is the set with the single element, $\mathrm{Ker}\ \phi$ (considered as the coset of e). Therefore, β is one to one.

67. A homomorphism $\phi: G \to G'$ is *onto* if and only if $\mathrm{Im}\ \phi = G'$ **(11)**. A homomorphism which is onto is called an *epimorphism*.

Proposition. *Every group homomorphism* $\phi: G \to G'$ *can be factored in the form* $\phi = \alpha\beta\gamma$, *where*

(1) α: Im $\phi \to G'$ *is a monomorphism,*
(2) β: $G/\text{Ker } \phi \to \text{Im } \phi$ *is an isomorphism,*
(3) γ: $G \to G/\text{Ker } \phi$ *is an epimorphism.*

Proof. The mapping α is simply the inclusion of the subset Im ϕ into G': it assigns to the element $x \in \text{Im } \phi$, the same element $x \in G'$; α is clearly one to one, a monomorphism. The mapping β is the isomorphism discussed in **66**. The mapping γ is given by $\gamma(a) = a(\text{Ker } \phi)$—that is, γ assigns to the element $a \in G$, the coset $a(\text{Ker } \phi)$, which belongs to $G/\text{Ker } \phi$. Since every coset of Ker ϕ is the coset of each of its elements, the mapping γ is onto. It is clear that γ is a homomorphism:

$$\gamma(ab) = (ab)(\text{Ker } \phi) = (a(\text{Ker } \phi))(b(\text{Ker } \phi)) = (\gamma a)(\gamma b).$$

Thus, γ is an epimorphism.

We can summarize the content of this proposition in Figure 4.

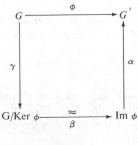

Figure 4

67α. Let **R** denote the group of real numbers under addition and **C*** the group of nonzero complex numbers under multiplication. Decompose the homomorphism ϕ: **R** $\to$ **C*** given by $\phi x = e^{2\pi i x}$ in the manner of the proposition above.

67β. For any group G let ϕ: $G \to \mathscr{A}(G)$ be given by $\phi g = \alpha_{g^{-1}}$ (**64**). Decompose ϕ in the manner of the proposition above.

67γ. Let ϕ: **Z** $\to$ **Z**$_m$ be the homomorphism given by $\phi(k) = [nk]_m$. Decompose ϕ as above.

67δ. Decompose the endomorphism ϕ: $K \to K$ of the circle group K given by $\phi z = z^n$.

68.　***The First Isomorphism Theorem.***　*If H and N are normal subgroups of a group G, and $N \subset H$, then H/N is a normal subgroup of G/N and there is an isomorphism*

$$(G/N)/(H/N) \approx G/H.$$

Proof.　We define a mapping $\phi: G/N \to G/H$ by $\phi(aN) = aH$. Clearly ϕ is well defined; furthermore ϕ is a homomorphism, since

$$\phi(aNbN) = \phi(abN) = abH = (aH)(bH) = \phi(aN)\phi(bN).$$

Now

$$\mathrm{Ker}\ \phi = \{aN \in G/N \mid \phi(aN) = H\} = \{aN \in G/N \mid a \in H\} = H/N.$$

Therefore by **65**, H/N is a normal subgroup of G/N. Furthermore, ϕ is onto (why?), and by **66**,

$$(G/N)/(H/N) = (G/N)/\mathrm{Ker}\ \phi \approx \mathrm{Im}\ \phi = G/H.$$

69.　***The Second Isomorphism Theorem.***　*If H and N are subgroups of a group G, and N is normal in G, then there is an isomorphism of groups,*

$$HN/N \approx H/(H \cap N).$$

Proof.　Tacit in the statement of this theorem are the statements: (1) $H \cap N$ is a normal subgroup of H; (2) HN is a subgroup of G; and (3) N is a normal subgroup of HN. We leave the proofs of these statements to the reader. (**46δ–46ε.**)
We define a mapping $\phi: HN/N \to H/(H \cap N)$ by

$$\phi(hN) = h(H \cap N).$$

(Note that $hnN = hN$—that is, for any element $hn \in HN$, the coset modulo N of hn is the same as that of h.) We must verify that

(1) ϕ *is well defined*: suppose $hN = h'N$; then $h' = hn$ for $n \in N$, and $n = h^{-1}h' \in H \cap N$; and consequently, $h(H \cap N) = h'(H \cap N)$.
(2) ϕ *is onto*: for any $h(H \cap N) \in H/(H \cap N)$, we have $h(H \cap N) = \phi(hN)$.
(3) ϕ *is one to one*: $\mathrm{Ker}\ \phi$ is the set of cosets $hN \in HN/N$ for which $h(H \cap N) = H \cap N$, which occurs only when $h \in H \cap N \subset N$ and $hN = N$.

Remark.　Since $HN = NH$, we also have $NH/N \approx H/(H \cap N)$.

70. **The Third Isomorphism Theorem.** (*Zassenhaus.*) *If H_1 and H_2 are subgroups of a group G and if N_1 and N_2 are normal subgroups of H_1 and H_2, respectively, then there are isomorphisms*

$$\frac{N_1(H_1 \cap H_2)}{N_1(H_1 \cap N_2)} \approx \frac{H_1 \cap H_2}{(H_1 \cap N_2)(N_1 \cap H_2)} \approx \frac{N_2(H_1 \cap H_2)}{N_2(N_1 \cap H_2)}.$$

Proof. Because of symmetry, only the first isomorphism need be proved. We leave it to the reader to verify that the requisite subgroups are normal. (See exercise **46v.**) Setting $H = H_1 \cap H_2$ and $N = N_1(H_1 \cap N_2)$, we apply **69** to obtain isomorphisms

$$\frac{N_1(H_1 \cap N_2)(H_1 \cap H_2)}{N_1(H_1 \cap N_2)} = \frac{NH}{N} \approx \frac{H}{H \cap N} = \frac{H_1 \cap H_2}{H_1 \cap H_2 \cap N_1(H_1 \cap N_2)}.$$

Exercise **35θ** states that *if X, Y, and Z are subgroups of a group and $Y \subset X$, then $X \cap YZ = Y(X \cap Z)$.* This implies:

  (1) $N_1(H_1 \cap N_2)(H_1 \cap H_2) = N_1(H_1 \cap H_2)$;
     (take $X = H_1$, $Y = H_1 \cap N_2$, $Z = H_2$);
  (2) $H_1 \cap H_2 \cap N_1(H_1 \cap N_2) = (H_1 \cap N_2)(N_1 \cap H_2)$;
     (take $X = H_1 \cap H_2$, $Y = H_1 \cap N_2$, $Z = N_1$).

The proof is now complete.

---

71. A *normal series* for a finite group G is a sequence of subgroups of G,

$$\{e\} = G_0 \subset G_1 \subset \cdots \subset G_n = G,$$

such that G_{i-1} is a proper normal subgroup of G_i for $i = 1, 2, \ldots, n$. The *factors* of a normal series are the quotient groups $G_1/G_0, G_2/G_1, \ldots, G_n/G_{n-1}$. A *refinement* of a normal series is a normal series which contains all the subgroups of the original normal series (and perhaps more). A refinement which is not identical with the original series is called a *proper refinement*.

72. **The Schreier-Zassenhaus Theorem.** *Two normal series for a finite group have refinements of equal length whose factor groups are isomorphic.*

Proof.   Suppose that

$$\{e\} = G_0 \subset G_1 \subset \cdots \subset G_n = G \tag{1}$$

and

$$\{e\} = H_0 \subset H_1 \subset \cdots \subset H_m = G \tag{2}$$

are two normal series for the finite group G. We form a new series of subgroups of G,

$$\{e\} = \hat{G}_0 \subset \hat{G}_1 \subset \cdots \subset \hat{G}_{nm} = G, \tag{3}$$

by setting $\hat{G}_k = G_q(G_{q+1} \cap H_r)$ for $k = qm + r$, where $0 \leq q < n$ and $0 \leq r \leq m$. Note that

$$\hat{G}_{qm} = G_q(G_{q+1} \cap H_0) = G_{q-1}(G_q \cap H_m) = G_q.$$

Thus, we see that $\hat{G}_k$ is well defined and that each group of the original series (1) occurs in series (3). Furthermore, each group $\hat{G}_k$ is clearly a normal subgroup of its successor $\hat{G}_{k+1}$. However, series (3) need not be a normal series—we may have $\hat{G}_k = \hat{G}_{k+1}$ for some values of k.

Similarly, we form another new series of subgroups of G,

$$\{e\} = \hat{H}_0 \subset \hat{H}_1 \subset \cdots \subset \hat{H}_{nm} = G, \tag{4}$$

by setting $\hat{H}_k = H_q(H_{q+1} \cap G_r)$ for $k = qn + r$, where $0 \leq q < m$ and $0 \leq r \leq n$. Remarks similar to those about series (3) apply to series (4).

Now we see that series (3) and (4) have isomorphic factors. In fact for $k = um + v$ and $l = vn + u$, we have by (**70**)

$$\frac{\hat{G}_{k+1}}{\hat{G}_k} = \frac{G_u(G_{u+1} \cap H_{v+1})}{G_u(G_{u+1} \cap H_v)} \approx \frac{H_v(H_{v+1} \cap G_{u+1})}{H_v(H_{v+1} \cap G_u)} = \frac{\hat{H}_{l+1}}{\hat{H}_l}.$$

Finally, we obtain refinements of the normal series (1) and (2) by eliminating the redundancies from series (3) and (4). Since (3) and (4) have isomorphic factors, it follows that they have the same number of redundancies. Consequently, the refinements of (1) and (2) obtained from them have the same length and isomorphic factors.

73.   A *composition series* for a finite group is a normal series which has no proper refinements. The following theorem is an almost immediate consequence of the Schreier-Zassenhaus theorem.

The Jordan-Hölder Theorem.   *Two composition series for a finite group have the same length and isomorphic factors.*

Proof. By the preceding theorem two composition series have refinements of equal length with isomorphic factors. However, the refinements must be identical with the original series, which by hypothesis have no proper refinements.

73α. Show that every finite group actually has a composition series.

73β. Define the *length* $l(G)$ of a finite group G to be the length (number of factors) of a composition series for G. Show that $H \lhd G$ implies that

$$l(G) = l(H) + l(G/H).$$

73γ. Let G be an abelian group of order $n = p_1^{v_1} p_2^{v_2} \cdots p_k^{v_k}$ where each p_i is prime. Show that $l(G) = v_1 + v_2 + \cdots + v_k$.

74. A group is *simple* if it has for normal subgroups only itself and the trivial group. For example, groups of prime order are necessarily simple, and there are many others. Our interest here in simple groups is due to the following result.

Proposition. *A normal series is a composition series if and only if each factor group is a simple group.*

Proof. Suppose that G is a finite group and that

$$\{e\} = G_0 \subset G_1 \subset \cdots \subset G_n = G \tag{*}$$

is a normal series for G. If $(*)$ is not a composition series, then we can obtain a proper refinement of $(*)$ by inserting a new group G' into the series at some point, say $G_k \subset G' \subset G_{k+1}$. It follows that G'/G_k is a normal subgroup of G_{k+1}/G_k and that G_{k+1}/G_k is not a simple group. On the other hand, if G_{k+1}/G_k is not simple for some k, then there is a normal subgroup G'',

$$\{e\} \subset G'' \subset G_{k+1}/G_k.$$

It follows that $G'' = G'/G_k$, where $G_k \subset G' \subset G_{k+1}$, and G' is a normal subgroup of G_{k+1}. (Why?) Then $(*)$ has a proper refinement and is not a composition series.

Corollary. *A group whose order is a power of a prime p has a composition series in which each factor is cyclic of order p.*

This is immediate from (**52**) and the proposition above.

74α. Construct a composition series for the dihedral group D_6 (**35ζ**).

74β.   Construct composition series for the groups Z_8, D_4, and Q (**46ι**).

75.   *Solvable Groups.* A finite group is *solvable* if it has a composition series in which each factor is a cyclic group. Since these factors must also be simple groups (**74**), they must all have prime order. Solvable groups are connected with the solvability of equations in radicals (**139–149**), which explains the unexpected terminology. We already know that every group of prime power order is solvable. We shall use the following results in Chapter 4.

Theorem.   *Let H be a proper normal subgroup of a group G. Then G is solvable if and only if H and G/H are solvable.*

Proof.   The normal series $\{e\} \subset H \subset G$ may be refined to a composition series for G,

$$\{e\} = G_0 \subset G_1 \subset \cdots \subset G_n = G.$$

Suppose that $G_k = H$. Then composition series for H and G/H are given by

$$\{e\} = G_0 \subset G_1 \subset \cdots \subset G_k = H$$

and

$$\{e\} = G_k/H \subset G_{k+1}/H \subset \cdots \subset G_n/H = G/H.$$

Furthermore, by the first isomorphism theorem (**68**) we have for $i \geq k$,

$$(G_{i+1}/H)/(G_i/H) \approx G_{i+1}/G_i.$$

Thus, each factor of the composition series for G is a factor of either the composition series for H or that for G/H. Now the theorem follows immediately.

Corollary.   *Every finite abelian group is solvable.*

Proof.   By induction on the order. The solvability of groups of orders 1, 2, and 3 is clear. As induction hypothesis we assume the solvability of all abelian groups with order below n. Let G be an abelian group of order n. There is some prime p which divides n. By Cauchy's theorem (**55**), G has an element of order p. Since G is abelian, this element generates a normal subgroup H of order p. If $p = n$ and $H = G$, we are finished because a group of prime order is clearly solvable. If $p < n$, then H and G/H (which has order n/p) are solvable and applying the theorem, G is solvable. This completes the induction step and the proof of the corollary.

75α. Prove that any subgroup of a solvable group is solvable.

75β. Prove that a group is solvable if it has a normal series whose factor groups are solvable.

75γ. Prove that a group with order below 60 is solvable. (See **59η**.)

75δ. Prove that a direct product of solvable groups is solvable.

75ε. Prove that a group is solvable which has just one p-Sylow subgroup for each prime p dividing its order.

he Symmetric Groups

The symmetric groups (**30**) are of such great importance in the Galois theory that we make a special study of their properties.

76. Let X denote a finite set. A *permutation* of X is a one-to-one onto mapping from X to X. The set $\mathscr{A}(X)$ of all permutations of X is a group in a natural way: if $S, T \in \mathscr{A}(X)$, then $ST \in \mathscr{A}(X)$ is the composite mapping, given by $(ST)x = S(Tx)$ for $x \in X$; the inverse of $S \in \mathscr{A}(X)$ is just the inverse mapping S^{-1}. A subgroup of $\mathscr{A}(X)$ will be called a *group of permutations of X*.

A permutation group is a special kind of transformation group (**53**). If G is a group of permutations of the finite set X, then the action of G on X is given by $g * x = g(x)$. This action satisfies:

(1) $g * (h * x) = (gh) * x$ for all $g, h \in G$ and all $x \in X$;
(2) $e * x = x$ for all $x \in X$;
(3) if $g * x = x$ for all $x \in X$, then $g = e$.

Only conditions (1) and (2) are required of transformation groups in general. An action of G on X which satisfies (3) is called *effective*, or alternatively, G is said to act *effectively*. It is clear that we could have made the definition: *a permutation group is a group which acts effectively on a finite set*. (See **53γ**.)

77. The structure of the full group of permutations $\mathscr{A}(X)$ of a finite set X is completely determined by the number of elements of X. To be precise, a one-to-one correspondence $\phi: X \to Y$ induces an isomorphism $\Phi: \mathscr{A}(X) \to \mathscr{A}(Y)$ given by $\Phi(S) = \phi S \phi^{-1}$. Suppose now that X is a set with n elements and that $\omega: X \to N_n$ is a one-to-one correspondence of X with the set $N_n = \{1, 2, \ldots, n\}$. (We might call ω an *ordering* of X.) Then $\omega: X \to N_n$ induces an isomorphism $\Omega: \mathscr{A}(X) \to \mathscr{A}(N_n)$. However, $\mathscr{A}(N_n)$ is simply S_n, the symmetric group on n letters, defined in **30**. In other words we have: *if X is a set with exactly n elements, then $\mathscr{A}(X)$ is isomorphic to S_n, the symmetric group on n letters.*

As a consequence, the study of permutation groups is reduced to the study of the symmetric groups and their subgroups. Every finite group can be viewed as a permutation group of its own set of elements and, consequently, is isomorphic with a subgroup of S_n where n is the order of the group. These observations indicate the significance of the symmetric groups.

77α. Show that every group may be considered as a group of permutations of its underlying set. (This is known as *Cayley's theorem* after the English mathematician Arthur Cayley (1821–1895), who was the first to consider abstract groups.)

78. Let $\pi: N_n \to N_n$ be an element of S_n, and let π_k denote $\pi(k)$. One way of expressing the permutation π is by a *tableau*

$$\pi = \begin{pmatrix} 1 & 2 & \cdots & n \\ \pi_1 & \pi_2 & \cdots & \pi_n \end{pmatrix}.$$

Clearly the order of elements in the top row is immaterial. Note that the inverse of π is

$$\pi^{-1} = \begin{pmatrix} \pi_1 & \pi_2 & \cdots & \pi_n \\ 1 & 2 & \cdots & n \end{pmatrix}.$$

This notation, used by Cauchy in his early studies of permutation groups, is needlessly complex. Each element of N_n appears twice, and no advantage is taken of the order in which the elements are written. We shall develop a more efficient notation in which every permutation is written as a product of "cycles" in a unique way.

The tableau notation makes it clear that *the order of S_n is $n!$*. The element π_1 may be chosen in n ways; once π_1 is chosen, there are $n - 1$ possibilities for π_2; when $\pi_1, \pi_2, \ldots, \pi_i$ have been chosen, there remain $n - i$ possible ways to choose π_{i+1}. Thus, there are $n \cdot (n - 1) \cdots 2 \cdot 1 = n!$ ways in which the bottom row of the tableau may be chosen, and thus there are $n!$ elements of S_n.

78α. Determine the number of permutations π which leave no element fixed, that is, for which $\pi_k \neq k$ for $k = 1, 2, \ldots, n$. (This is a famous, but difficult, problem which was first solved by Nicolas Bernoulli (1687–1759), and later, independently, by Euler.)

79. Let $a_1, a_2, \ldots, a_k \in N_n$ be distinct integers. We shall denote by $(a_1, a_2, \ldots, a_k)$ the permutation

$$\begin{pmatrix} a_1 & a_2 & \cdots & a_k & \cdots & i & \cdots \\ a_2 & a_3 & \cdots & a_1 & \cdots & i & \cdots \end{pmatrix}$$

which carries a_1 to a_2, a_2 to a_3, $\ldots$, and a_k to a_1, leaving all the other elements of N_n fixed. We call $(a_1, a_2, \ldots, a_k)$ a *cyclic permutation of order k* or a *k-cycle*. This notation is almost too efficient: $(a_1, a_2, \ldots, a_k)$ can denote an element of any one of the groups S_n for which $n \geq k$.

A cyclic permutation of order 2, (a_1, a_2), simply interchanges a_1 and a_2 and is called a *transposition*.

Two cyclic permutations $(a_1, a_2, \ldots, a_k)$ and $(b_1, b_2, \ldots, b_l)$ are *disjoint* if they have no entries in common. *Disjoint cyclic permutations commute*, that is

$$(a_1, a_2, \ldots, a_k)(b_1, b_2, \ldots, b_l) = (b_1, b_2, \ldots, b_l)(a_1, a_2, \ldots, a_k).$$

However the groups S_n are not abelian for $n > 2$.

79α. Compute the number of distinct k-cycles in S_n.

79β. Show that if $\pi \in S_n$ and $n > 2$, then there exists a transposition τ such that $\tau\pi \neq \pi\tau$ (unless, of course, π is the identity element of S_n). This shows that S_n has trivial center for $n > 2$.

79γ. Prove that disjoint cyclic permutations commute.

79δ. Show that S_n contains $\binom{n}{k}$ subgroups isomorphic to $S_k \times S_{n-k}$, all of which are conjugates. $\left(\text{See } \mathbf{20\gamma} \text{ for definition of the binomial coefficient } \binom{n}{k}.\right)$

80. Theorem. *Every permutation of n letters is the product of disjoint cyclic permutations in exactly one way (except for order of the factors).*

Proof. Let $\pi \in S_n$. We shall denote by H the cyclic subgroup of S_n generated by π. H acts on the set $N_n = \{1, 2, 3, \ldots, n\}$ dividing it into disjoint orbits, $X_1, X_2, \ldots, X_r$. In other words, two elements i and j of N_n belong to the same orbit if and only if $j = \pi^k(i)$ for some power π^k of π. In any orbit X_k we may list elements in order

$$a_{k1}, a_{k2}, \ldots, a_{ks_k},$$

so that $a_{k\,i+1} = \pi(a_{ki})$ and $a_{k1} = \pi(a_{ks_k})$. (Why is this possible?) We let α_k denote the cyclic permutation $(a_{k1}, a_{k2}, \ldots, a_{ks_k})$. We claim that $\pi = \alpha_1\alpha_2 \ldots \alpha_r$.

To prove this we need only show that π and $\alpha_1\alpha_2 \cdots \alpha_r$ have the same effect on any element $x \in N_n$. If $x \in X_k$, then $\alpha_i(x) = x$ for $i \neq k$ and $\alpha_k(x) = \pi(x)$. Therefore,

$$(\alpha_1\alpha_2 \cdots \alpha_r)(x) = \alpha_k(x) = \pi(x).$$

Finally, the expression $\pi = \alpha_1\alpha_2 \cdots \alpha_r$ is clearly unique except for the order of the α_i's.

Note that we may include or exclude factors of the form $\alpha_k = (m)$ since every 1-cycle is the identity.

In practice it is a simple matter to express a permutation as the product of disjoint cyclic permutations. For example

$$\begin{pmatrix} 1 & 2 & 3 & 4 & 5 & 6 & 7 \\ 5 & 6 & 1 & 7 & 3 & 2 & 4 \end{pmatrix} = (1, 5, 3)(2, 6)(4, 7).$$

Corollary. *If* $\alpha_1, \alpha_2, \ldots, \alpha_s \in S_n$ *are disjoint cyclic permutations, then the order of* $\alpha_1\alpha_2 \cdots \alpha_s$ *is the least common multiple of the orders of the factors.*

Proof. Let k_i denote the order of α_i, and let k be the least common multiple of the k_i. Since the α_i's commute, we have

$$(\alpha_1\alpha_2 \cdots \alpha_s)^k = \alpha_1^k \alpha_2^k \cdots \alpha_s^k = e$$

so that $o(\alpha_1\alpha_2 \cdots \alpha_s)$ divides k. Since the α_i are disjoint, it follows that $(\alpha_1\alpha_2 \cdots \alpha_s)^l = e$ implies $\alpha_i^l = e$ for each i. Then $k_i \mid l$ for each i, and thus $k \mid l$. In particular $k \mid o(\alpha_1\alpha_2 \cdots \alpha_s)$, and therefore $o(\alpha_1\alpha_2 \cdots \alpha_s) = k$.

Corollary. *Every permutation is a product of transpositions.*

Proof. It is enough to show that every cyclic permutation is a product of transpositions. This is easy because

$$(a_1, a_2, \ldots, a_k) = (a_1, a_k)(a_1, a_{k-1}) \cdots (a_1, a_2).$$

80α. Let $\pi \in S_n$ be written as a product $\alpha_1\alpha_2 \cdots \alpha_n$ of disjoint cycles, where for convenience we assume that

$$o(\alpha_1) \geq o(\alpha_2) \geq \cdots \geq o(\alpha_n).$$

We shall call the decreasing sequence of natural numbers $o(\alpha_1)$, $o(\alpha_2)$, ..., $o(\alpha_n)$ the *form* of π. Show that two elements of S_n are conjugate if and only if they have the same form.

80β. Compute the number of elements which have the form $k_1, k_2, \ldots, k_m$.

80γ. Show that S_n is generated by the transpositions

$$(1, 2), (2, 3), \ldots, (n - 1, n).$$

80δ. Show that S_n is generated by the cycles $(1, 2)$ and $(1, 2, \ldots, n)$.

81. *Even and Odd Permutations.* Let $\mathscr{P}_n$ denote the polynomial of n variables $x_1, x_2, \ldots, x_n$ which is the product of all the factors $x_i - x_j$ with $i < j$; that is,

$$\mathscr{P}_n(x_1, x_2, \ldots, x_n) = \prod_{i<j} (x_i - x_j).$$

The symmetric group S_n acts on the polynomial $\mathscr{P}_n$ by permuting the variables. For $\pi \in S_n$ we have

$$\mathscr{P}_n(x_{\pi(1)}, x_{\pi(2)}, \ldots, x_{\pi(n)}) = (sgn \, \pi)\mathscr{P}_n(x_1, x_2, \ldots, x_n),$$

where $sgn \, \pi = \pm 1$. If the sign is positive, then π is called an *even permutation*; if the sign is negative, then π is called an *odd permutation*. (*Sgn* is an abbreviation of the Latin word *signum*.)

It is not difficult to see that $sgn(\pi\sigma) = (sgn \, \pi)(sgn \, \sigma)$. This means that:

the product of two even or two odd permutations is even,
the product of an even and an odd permutation is odd.

It follows that the set of even permutations is a subgroup of S_n, called the *alternating group on n letters* and customarily denoted A_n. We may regard $sgn : S_n \to K_2$ as a homomorphism from S_n to $K_2 = \{\pm 1\}$, the group of square roots of unity (**44**). This shows that A_n *is a normal subgroup of* S_n and the quotient group S_n/A_n is isomorphic to K_2.

It is immediate that $o(A_n) = n!/2$.

81α. Determine the sign of a k-cycle in terms of k.

82. *Proposition. Every even permutation of n letters, $n \geq 3$, is the product of cyclic permutations of order 3.*

Proof. There are no cyclic permutations of order 3 in S_n for $n = 1$ or $n = 2$, but the identity is always an even permutation. Thus, the proposition is false for $n < 3$. Our proof will be inductive, beginning with $n = 3$.

The even permutations of 3 letters are the identity e and the 3-cycles $(1, 2, 3)$ and $(1, 3, 2)$. Thus, we have easily disposed of the case $n = 3$.

Now suppose the proposition proved for even permutations of less than n letters, and suppose $\pi \in A_n$. The permutation $\sigma = (\pi_n, n, i) \cdot \pi$, where $\pi_i = n$, satisfies

$$\sigma(n) = (\pi_n, n, i) \cdot \pi(n) = (\pi_n, n, i)(\pi_n) = n$$

and is even. Since σ leaves n fixed and is even, σ may be considered as an even permutation of the letters $1, 2, \ldots, n - 1$. By the inductive hypothesis σ is the product of 3-cycles, say $\sigma = \alpha_1\alpha_2 \cdots \alpha_s$. Setting $\alpha_0 = (\pi_n, i, n)$, we have

$$\alpha_0\alpha_1\alpha_2 \cdots \alpha_s = \alpha_0\sigma = (\pi_n, i, n)(\pi_n, n, i) \cdot \pi = \pi,$$

and we have expressed π as a product of 3-cycles.

82α.   Show that the alternating group A_n is generated by the 3-cycles $(1, i, n)$ for $i = 2, 3, \ldots, n - 1$.

83.   Theorem.   *The alternating group A_n is simple except for $n = 4$.*

Proof.   Recall that a group is simple if it has only itself and the trivial group as normal subgroups. For $n < 4$ the order of A_n is either 1 or 3, and A_n is obviously simple. The major part of the proof is the case $n > 4$.

Let N be a nontrivial normal subgroup of A_n for $n > 4$. We must show that $N = A_n$. The first step is to see that N contains a 3-cycle.

Let $\alpha \neq e$ be an element of N which leaves fixed as many elements of $\mathbf{N}_n$ as possible. As guaranteed by **80**, let

$$\alpha = \alpha_1\alpha_2 \cdots \alpha_s$$

where the α_i are disjoint cycles, which we may assume are given in order of decreasing length. Renumbering if necessary, we may assume that

$$\alpha_1 = (1, 2, \ldots, k)$$

and, when $s > 1$, that

$$\alpha_2 = (k + 1, k + 2, \ldots, l).$$

We distinguish several cases.

Case 1. α *moves each of the numbers* 1, 2, 3, 4, 5. (This occurs when $s > 2$, when $s = 2$ and $\alpha = (1, 2, \ldots, k)(k + 1, k + 2, \ldots, l)$ with $l > 4$, or when $s = 1$ and $\alpha = \alpha_1 = (1, 2, \ldots, k)$ for $k > 4$.) Setting $\beta = (3, 4, 5)$, the element $\beta^{-1}\alpha^{-1}\beta$ belongs to the normal subgroup N, and thus $\beta^{-1}\alpha^{-1}\beta\alpha \in N$. However, it is easily checked that the permutation $\beta^{-1}\alpha^{-1}\beta\alpha$ leaves the number 1 fixed in addition to leaving fixed all the elements fixed by α. This contradicts the choice of α, and case 1 is impossible.

Case 2. α *moves the numbers* 1, 2, 3, 4 *and no others.* (This occurs only when $\alpha = (1, 2)(3, 4)$, since $(1, 2, 3, 4)$ is an odd permutation.) Again we set $\beta = (3, 4, 5)$ and argue that the element $\beta^{-1}\alpha^{-1}\beta\alpha$ belongs to N. However, direct computation shows that $\beta^{-1}\alpha^{-1}\beta\alpha = (3, 4, 5) = \beta$. Thus, $\beta \in N$ and β moves fewer elements than α. This contradiction eliminates case 2.

Case 3. α *moves the numbers* 1, 2, 3 *and no others.* (This occurs only when $\alpha = (1, 2, 3)$.) There are no other cases now that the first and second are eliminated. Thus, we have shown that N contains a 3-cycle, which we may assume to be $(1, 2, 3)$.

It remains to show that N contains every 3-cycle. Choose an even permutation

$$\sigma = \begin{pmatrix} 1 & 2 & 3 & \cdots \\ i & j & k & \cdots \end{pmatrix}.$$

Then, $\sigma(1, 2, 3)\sigma^{-1} = (i, j, k)$ belongs to the normal subgroup N. Varying i, j, and k, we obtain all 3-cycles. Thus, N contains every 3-cycle, and in view of **82**, $N = A_n$, and we are finished.

Remark. The group A_4 is not simple: it contains a normal subgroup of order 4 containing the elements e, $(1, 2)(3, 4)$, $(1, 3)(2, 4)$, and $(1, 4)(2, 3)$.

83α. Verify that the set

$$N = \{e, \ (1, 2)(3, 4), \ (1, 3)(2, 4), \ (1, 4)(2, 3)\}$$

is a normal subgroup of A_4. Show that $K = \{e, (1, 2)(3, 4)\}$ is normal in N but not in A_4. (This shows that a normal subgroup of a normal subgroup need not be normal in the whole group.)

83β. Show that A_4 has no subgroup of order 6.

84. Theorem. *For $n > 4$, the symmetric group S_n is not solvable.*

Proof. Since the groups $K_2 \approx S_n/A_n$ and A_n (for $n > 4$) are simple, the normal series

$$\{e\} \subset A_n \subset S_n$$

is a composition series for S when $n > 4$. However, A_n is not abelian for $n > 3$. (For example, $(1, 2, 3)(2, 3, 4) = (1, 2)(3, 4)$ while $(2, 3, 4)(1, 2, 3) = (1, 3)(2, 4)$.) Consequently, A_n is not cyclic for $n > 3$. As a result S_n is not solvable for $n > 4$.

84α. Construct composition series for the groups S_2, S_3, and S_4, and verify that these groups are solvable.

84β. Show that for $n > 4$ the only normal subgroups of S_n are $\{e\}$, A_n, and S_n itself.

85. We have already remarked that a finite group can be viewed as a permutation group of its set of elements. We can improve this as follows.

Theorem. *If H is a subgroup of a finite group G and H contains no nontrivial normal subgroup of G, then G is isomorphic to a subgroup of $\mathscr{A}(G/H)$, the group of permutations of the set G/H.*

Proof. Define a homomorphism $\phi\colon G \to \mathscr{A}(G/H)$ by setting

$$\phi(g)(xH) = (gx)H \quad \text{all } x \in G.$$

Ker ϕ is a normal subgroup of G (**65**). An element g belongs to Ker ϕ if and only if $(gx)H = xH$ for all $x \in G$, or what is the same thing, $x^{-1}gx \in H$ for all $x \in G$. In other words, Ker ϕ is the intersection of H and all its conjugates. Thus, Ker $\phi \subset H$ and by hypothesis Ker ϕ must be trivial. It follows that $G \approx \operatorname{Im} \phi$.

Corollary. *For $n > 4$, A_n is the only proper subgroup of index less than n in S_n.*

Proof. It follows from **84β** that for $n > 4$, A_n is the only proper, nontrivial, normal subgroup of S_n. Suppose that H is a subgroup of S_n and $[S_n : H] < n$. If $[S_n : H] = 2$, then H is normal and $H = A_n$. On the other hand $[S_n : H] > 2$ implies $A_n \not\subset H$. Thus, the hypothesis of the theorem is satisfied, and S_n is isomorphic to a subgroup of $\mathscr{A}(S_n/H)$. However,

$$o(\mathscr{A}(S_n/H)) = [S_n : H]! < n! = o(S_n),$$

which is a contradiction.

85α. Let $f(x_1, x_2, \ldots, x_n)$ be a function of n variables, $n > 4$. Let v denote the number of distinct functions obtained when the variables $x_1, x_2, \ldots, x_n$ are permuted. Show that $v > 2$ implies $v \geq n$. (The general problem of what can be said about the number v is classic and one of the motivating ideas for group theory. Results of this type were given by many early group theorists, including LaGrange, Ruffini, Abel, Cauchy, and Galois.)

86. A subgroup H of S_n is *transitive* if for every pair of elements $i, j \in \mathbf{N}_n$ there is an element $\pi \in H$ such that $\pi(i) = j$. For example, the cycle $\alpha = (1, 2, \ldots, n)$ generates a transitive subgroup of S_n: the element α^{j-i} carries i to j. The following theorem about transitive subgroups will be needed to establish unsolvability of quintic equations in general.

Theorem. *Let H be a transitive subgroup of S_p where p is a prime number. If H contains a transposition, then $H = S_p$.*

Proof. We may assume without loss of generality that $(1, 2)$ is the transposition H contains. An equivalence relation on the set $\mathbf{N}_p = \{1, 2, \ldots, p\}$ is defined by $i \sim j$ if and only if the transposition $(i, j) \in H$. From the transitivity of H it follows that each equivalence class has the same number of elements; in fact, if $\phi \in H$ and $\phi_1 = \phi(1) = i$, then ϕ yields a one-to-one correspondence from the equivalence class of 1 to that of i since $(1, k) \in H$ if and only if

$$(i, \phi_k) = (\phi_1, \phi_k) = \phi \cdot (1, k) \cdot \phi^{-1} \in H.$$

The number s of elements in any equivalence class must divide the prime p, and thus $s = 1$ or $s = p$. However, the equivalence class of 1 contains at least the two elements 1 and 2. Consequently, there can be only one equivalence class which has p elements. In other words, H contains all the transpositions of S_p. Since every permutation is a product of transpositions (**80**), we have $H = S_p$.

Field Theory

Chapter **3**

A field is an algebraic structure in which the four rational operations, addition, subtraction, multiplication, and division, can be performed and in which these operations satisfy most of the familiar rules of arithmetical operations with numbers. In the formal definition of field structure, we assume only that addition and multiplication are given; subtraction and division are defined as the inverse operations. Division by 0 is automatically prohibited by the definition.

Field theory is the theoretical background for the theory of equations. It does not make sense to ask, for example, whether the equation $x^2 + x + 1 = 0$ is solvable, without specifying the field in which we want the solutions to lie. If we specify the field to be the set **R** of all real numbers, then the equation $x^2 + x + 1 = 0$ has no solutions, which is to say, there are no real numbers satisfying this equation. On the other hand there are complex numbers (the cube roots of unity, ω and ω^2) which do satisfy this equation in the field **C** of all complex numbers.

From an abstract viewpoint the theory of equations is just the study of field theory. In this chapter we present the basic field theory which is needed for Galois theory in the next chapter. To illustrate the depth of field theory, we take up the ancient problem of constructibility of geometric figures with straightedge and compass and prove that, in general, angles are not trisectable in this way.

87. A *field* is a set F with two operations (called *addition* and *multiplication*) which assign to each ordered pair (a, b) of elements from F, two elements of F, called their *sum*, $a + b$, and their *product*, ab, in such a way that:

(1) F is an abelian group under addition (with identity element 0);
(2) F^*, the set of nonzero elements of F, is an abelian group under multiplication;
(3) multiplication is distributive over addition; that is, for any three elements $a, b, c \in F$,

$$a(b + c) = ab + ac \qquad \text{and} \qquad (a + b)c = ac + bc.$$

As customary in the additive notation for abelian groups, we shall denote the additive inverse of $a \in F$ by $-a$. If $a \in F^*$, then a has a multiplicative inverse as well, and we denote this by a^{-1} or $1/a$. We always denote the multiplicative identity element (identity element of the group F^*) by 1.

87α. Show that $0a = 0 = a0$ for any element a of a field F.

87β. Show that $(-1)a = -a$ for any element a of a field F.

87γ. Let a and b be elements of a field F such that $ab = 0$. Show that $a = 0$ or $b = 0$.

87δ. If a and b are elements of a field F and $b \neq 0$, let a/b denote ab^{-1}. Show that when $a \neq 0$, $1/(a/b) = b/a$. For $a, c \in F$ and $b, d \in F^*$ prove the rule for addition of fractions:

$$\frac{a}{b} + \frac{c}{d} = \frac{ad + bc}{bd}.$$

87ε. Construct a field with four elements.

87ζ. Let F be a field and let $E = F \times F$. Define addition and multiplication in E by the rules:

$$(a, b) + (c, d) = (a + c, b + d)$$

and

$$(a, b)(c, d) = (ac - bd, ad + bc).$$

Determine conditions on F under which E (with these operations) is a field.

We extend the notion of linear independence to infinite sets as follows: an infinite subset of E is linearly independent over F if each one of its finite subsets is linearly independent over F.

92.   Suppose that E is a vector space over F. A set S of elements of E is a *spanning set* for E over F if every element of E can be written as a linear combination,

$$c_1\sigma_1 + c_2\sigma_2 + \cdots + c_n\sigma_n,$$

of elements $\sigma_1, \sigma_2, \ldots, \sigma_n \in S$ and coefficients $c_1, c_2, \ldots, c_n \in F$.

If there exists a finite spanning set for E over F, then E is called a *finite dimensional vector space* over F.

92α.   Let S be any subset of a vector space E over F. Show that the set of vectors E' of E which can be written as linear combinations of vectors in S is a vector space over F. (E' is said to be a subspace of E. See **92β**.)

92β.   A subset E' of a vector space E over F is a *subspace* of E' if every linear combination of vectors in E' belongs to E'. Show that a subspace of a vector space is again a vector space (over the same field).

92γ.   A *linear transformation* from a vector space E over a field F to a vector space E' over F is a mapping $T: E \to E'$ such that

$$T(\alpha + \beta) = (T\alpha) + (T\beta) \quad \text{and} \quad T(c\alpha) = c(T\alpha)$$

for all vectors α and β and every scalar c in F. Show that the sets

$$\text{Ker } T = \{\alpha \in E \,|\, T\alpha = 0\} \quad \text{and} \quad \text{Im } T = \{\alpha' \in E' \,|\, \alpha' = T\alpha, \alpha \in E\}$$

are subspaces of E and E', respectively.

93.   Again suppose that E is a vector space over F. A *basis* for E over F is a minimal spanning set for E over F. Explicitly, a set B of elements of E is a basis for E over F if

(1) B is a spanning set for E over F,
(2)   no proper subset of B spans E over F.

Proposition.   *A basis is linearly independent.*

Proof.   Let $\{\beta_1, \beta_2, \ldots, \beta_n\}$ be a finite subset of a basis B for E over F. We shall see that if $\{\beta_1, \beta_2, \ldots, \beta_n\}$ were linearly dependent, B could not be minimal. Suppose that

$$c_1\beta_1 + c_2\beta_2 + \cdots + c_n\beta_n = 0 \tag{*}$$

were a linear relation among the β_i's with $c_1, c_2, \ldots, c_n \in F$, not all zero. We may suppose, without loss of generality, that $c_1 \neq 0$. Then it would follow from (*) that

$$\beta_1 = -\left(\frac{c_2}{c_1}\right)\beta_2 - \left(\frac{c_3}{c_1}\right)\beta_3 - \cdots - \left(\frac{c_n}{c_1}\right)\beta_n,$$

in other words, that β_1 is a linear combination of the elements $\beta_2, \beta_3, \ldots, \beta_n$ with coefficients in F. It is not hard to see that since B spans E over F, the set $B - \{\beta_1\}$ would also span E over F. However, this would show that B does not satisfy condition (2) and is not a basis. Therefore it must be true that every finite subset of B is linearly independent and that B itself is linearly independent.

Proposition. *A linearly independent spanning set is a basis.*

Proof. Suppose that B is a linearly independent spanning set for E over F. If B is not a basis, then some proper subset S of B also spans E over F. Choose $\beta \in B - S$. Since S spans E over F, there are elements $\sigma_1, \sigma_2, \ldots, \sigma_m \in S$ and coefficients $c_1, c_2, \ldots, c_m \in F$ such that

$$\beta = c_1\sigma_1 + c_2\sigma_2 + \cdots + c_m\sigma_m.$$

However, this implies that the set $\{\beta, \sigma_1, \sigma_2, \ldots, \sigma_m\}$, which is a finite subset of B, is linearly dependent over F. This contradicts the linear independence of B. Thus, B must be a basis.

94. Proposition. *If B is a basis for E over F, then every element of E may be written uniquely as a linear combination of elements of B with coefficients in F.*

Proof. We have only to prove uniqueness since B spans E over F. Suppose that $\beta \in E$ can be written in two ways as a linear combination of elements of B, say

$$\beta = c_1\beta_1 + c_2\beta_2 + \cdots + c_n\beta_n,$$
$$\beta = d_1\beta_1 + d_2\beta_2 + \cdots + d_n\beta_n,$$

where we assume without loss of generality that the same elements, $\beta_1, \beta_2, \ldots, \beta_n$, of B are involved in the linear combinations. Now subtracting the two expressions above, we have

$$(c_1 - d_1)\beta_1 + (c_2 - d_2)\beta_2 + \cdots + (c_n - d_n)\beta_n = 0.$$

Linear independence of the set $\{\beta_1, \beta_2, \ldots, \beta_n\}$ implies that $c_i = d_i$ for $i = 1, 2, \ldots, n$. Uniqueness is proved.

95. Proposition. *If E is a finite dimensional vector space over F, then every basis for E over F is finite and all such bases have the same number of elements.*

Proof. Since E is finite dimensional over F (**92**), there is a finite spanning set for E over F from which a finite basis can be selected. (How?) Let $B = \{\beta_1, \beta_2, \ldots, \beta_n\}$ be such a finite basis, and let A be any other basis.

Choose an element $\alpha_1 \in A$. Let α_1 be written as

$$\alpha_1 = c_1\beta_1 + c_2\beta_2 + \cdots + c_n\beta_n, \quad c_i \in F. \tag{1}$$

Since α_1 is a basis element and therefore nonzero, not all the coefficients c_i are zero. Without loss of generality we may suppose that $c_1 \neq 0$. Now we claim that the set $B_1 = \{\alpha_1, \beta_2, \ldots, \beta_n\}$, in which α_1 has replaced β_1, is again a basis for E over F. Since

$$\beta_1 = c_1^{-1}\alpha_1 - c_2\beta_2 - \cdots - c_n\beta_n, \tag{2}$$

every element of E can be written first in terms of the basis elements $\beta_1, \beta_2, \ldots, \beta_n$; then the right hand side of (2) can be substituted for β_1 yielding an expression in terms of the elements $\alpha_1, \beta_2, \ldots, \beta_n$. This shows that B_1 spans E over F. It remains to show that B_1 is linearly independent. Suppose now that

$$d_1\alpha_1 + d_2\beta_2 + \cdots + d_n\beta_n = 0, \quad d_i \in F, \tag{3}$$

is a linear relation among the elements of B_1. Substituting (1) into (3) gives a relation among the elements of B:

$$(d_1c_1)\beta_1 + (d_2 + d_1c_2)\beta_2 + \cdots + (d_n + d_1c_n)\beta_n = 0. \tag{4}$$

The linear independence of B implies that the coefficients in (4) are all zero, from which we conclude that $d_1 = 0$ (since $c_1 \neq 0$ by assumption above), and consequently, that $d_2 = d_3 = \cdots = d_n = 0$. We have shown that B_1 is a linearly independent spanning set for E over F and thus a basis by (**93**).

Next choose an element $\alpha_2 \in A$ which is not a scalar multiple of α_1. (If no such α_2 exists, it must be that $n = 1$ and the argument is finished.) We can write

$$\alpha_2 = c_1'\alpha_1 + c_2'\beta_2 + \cdots + c_n'\beta_n, \quad c_i' \in F.$$

Since α_2 is linearly independent of α_1, not all the coefficients $c_2', c_3', \ldots, c_n'$ are zero. Without loss of generality, we may assume that $c_2' \neq 0$. Now we claim that the set $B_2 = \{\alpha_1, \alpha_2, \beta_3, \ldots, \beta_n\}$ is a basis for E over F, which is proved by an argument like that above.

Continuing in this fashion, we arrive at a basis $B_n = \{\alpha_1, \alpha_2, \ldots, \alpha_n\}$ made up entirely of elements from A. Since A is a basis and minimal, it follows that $B_n = A$.

The number of elements in a basis for E over F is called the *dimension* of E over F and is denoted $[E: F]$.

95α. Prove that a subspace E' of a finite dimensional vector space E over F is again finite dimensional and that $[E':F] \leq [E: F]$.

95β. Let E' be a subspace of a vector space E over F. An equivalence relation on E is defined by $\alpha \equiv \beta \bmod E'$ if and only if $\alpha - \beta \in E'$, and we denote the quotient set of this equivalence relation by E/E'. Show that E/E' is a vector space over F. Show that E finite dimensional implies that E/E' is finite dimensional.

95γ. With the same hypothesis as in **95β**, show that when E is finite dimensional the dimension of E is the sum of the dimensions of E' and E/E'.

:tension Fields

---

96. A field E is called an *extension (field)* of a field F if F is a subfield of E. This additional terminology seems superfluous, and technically it is. It reveals, however, a difference in modes of thought between field theory and group theory. In group theory we are often interested in determining the substructure of a group whereas in field theory we are more interested in what superstructures a field can support. Frequently we shall extend a field by adjoining to it additional elements.

Proposition. *An extension field E of a field F is a vector space over F.*

Proof. Clearly, E is an abelian group under addition. Scalar multiplication of an element $c \in F$ and an element $\alpha \in E$ is defined as the product $c\alpha$ where *both* c and α are considered as elements of E. Now the four properties required of scalar multiplication (**90**) are immediate consequences of E being a field.

When an extension field E of a field F is a finite dimensional vector space over F, we shall refer to E as a *finite extension* of F and to the dimension $[E: F]$ as the *degree* of E over F.

Finally, we note that a sequence of extension fields

$$F_0 \subset F_1 \subset \cdots \subset F_n$$

is called a *tower* of fields, and F_0 is called the *ground field*.

96α. Show that the degree of $\mathbf{Q}(\sqrt{2})$ over $\mathbf{Q}$ is 2 (**88β**).

96β. Show that $[\mathbf{C} : \mathbf{R}] = 2$.

96γ. Let $\omega = e^{2\pi i/3} = -\frac{1}{2} + \frac{1}{2}\sqrt{-3}$. Show that the set

$$\mathbf{Q}(\omega) = \{z \in \mathbf{C} \mid z = a + b\omega; \, a, b \in \mathbf{Q}\}$$

is an extension field of $\mathbf{Q}$ of degree 2.

96δ. Show that a finite field (that is, a field with a finite number of elements) of characteristic p (**89α**) has p^n elements for some n.

97. Proposition. *If D is a finite extension of E and E is a finite extension of F, then D is a finite extension of F. Furthermore,*

$$[D : F] = [D : E][E : F].$$

Proof. Let $A = \{\alpha_1, \alpha_2, \ldots, \alpha_m\}$ be a basis for E over F, and let $B = \{\beta_1, \beta_2, \ldots, \beta_n\}$ be a basis for D over E. We shall show that the set

$$C = \{\alpha_i \beta_j \mid 1 \le i \le m, 1 \le j \le n\}$$

is a basis for D over F.

(1) *C spans D over F.* Suppose $\gamma \in D$. Using the basis B, we write

$$\gamma = \gamma_1 \beta_1 + \gamma_2 \beta_2 + \cdots + \gamma_n \beta_n, \qquad \gamma_i \in E.$$

Each of the elements $\gamma_i \in E$, $i = 1, 2, \ldots, n$, can be written as

$$\gamma_i = c_{i1}\alpha_1 + c_{i2}\alpha_2 + \cdots + c_{im}\alpha_m, \qquad c_{ij} \in F.$$

Substituting these expressions into the one above yields

$$\gamma = \sum_{i=1}^{n} \sum_{j=1}^{m} c_{ij} \, \alpha_j \beta_i, \qquad c_{ij} \in F.$$

(2) *C is linearly independent.* Suppose that there is a linear relation among the elements of C with coefficients in F,

$$\sum_{i=1}^{n} \sum_{j=1}^{m} c_{ij}\alpha_j \beta_i = 0, \qquad c_{ij} \in F.$$

We may regard this as a linear relation among the elements of B with coefficients $\gamma_i = \sum_{j=1}^m c_{ij} \alpha_j$, which must be zero since B is a basis and linearly independent. On the other hand, the linear relations

$$\sum_{j=1}^m c_{ij} \alpha_j = 0, \qquad i = 1, 2, \ldots, m,$$

imply that all the coefficients c_{ij} are zero for

$$i = 1, 2, \ldots, n \quad \text{and} \quad j = 1, 2, \ldots, m.$$

We have shown that C is a linearly independent spanning set for D over F; hence it is a basis. Since C has a finite number of elements, it follows that D is a finite extension of F. Finally, we have

$$[D : F] = nm = [D : E][E : F].$$

ynomials

98. A *polynomial over a field F in the indeterminate x* is an expression of the form

$$c_0 + c_1 x + \cdots + c_n x^n,$$

where $c_0, c_1, \ldots, c_n$ are elements of F, called *coefficients* of the polynomial. Polynomials are completely determined by their coefficients, which is to say, two polynomials over F in x are equal if their corresponding coefficients are equal.

The phrase "an expression of the form" in a mathematical definition is hardly consonant with modern standards of rigor, and we shall eventually give a more precise treatment of polynomials (**156**).

We shall usually denote polynomials by a single letter such as f and write an equation such as

$$fx = c_0 + c_1 x + \cdots + c_n x^n$$

to specify the coefficients. Then, $f(2)$, $f(x^p)$, $f(y + 1)$, etc., will indicate the corresponding expressions in which 2, x^p, $y + 1$, etc. have been substituted for x. The largest number k for which $c_k \neq 0$ is called the *degree* of f (denoted $\deg f$), and c_k is called the *leading coefficient* of f. If all the coefficients of f are zero, we write $f = 0$ and do not assign a degree to f.

Polynomials are added and multiplied just as in elementary algebra, and we have

$$\deg(f + g) \le \max\{\deg f, \deg g\},$$

$$\deg(fg) = \deg f + \deg g,$$

whenever the polynomials involved are nonzero.

If f is a polynomial over F in x and $a \in F$, then fa is an element of F. The assignment $a \to fa$ defines a function $F \to F$, which we shall denote f. It may happen, however, that two distinct polynomials define the same function. For example, the polynomials x and x^p over $\mathbf{Z}_p$ have this property. (Why?)

$F[x]$ will denote the set of all polynomials in x over F. A polynomial of the form $fx = c$ is called a *constant polynomial* and will sometimes be identified with the corresponding element $c \in F$. In this way we may view F as a subset of $F[x]$.

98α.   A *rational function over a field F in the indeterminate x* is a (formal) quotient p/q of polynomials p and q over F. Two such quotients, p/q and r/s, are equal if and only if $ps = qr$ in $F[x]$. We denote the set of all rational functions of x over F by $F(x)$. We identify a rational function $p/1$ (where 1 is the constant polynomial $fx = 1$) with the polynomial p. Thus, $F[x]$ is identified with a subset of $F(x)$. (Sometimes polynomials are called *integral functions*.) Show that $F(x)$ is a field under the operations defined by

$$\frac{p}{q} + \frac{r}{s} = \frac{ps + qr}{qs}, \qquad \left(\frac{p}{q}\right)\left(\frac{r}{s}\right) = \frac{pr}{qs}.$$

99.   The Division Theorem for Polynomials.   *If f and g are polynomials over F and $g \ne 0$, then there exist over F unique polynomials q and r such that $f = qg + r$ and either $r = 0$ or $\deg r < \deg g$.*

Proof.   Let R denote the set of all polynomials over F which have the form $f - qg$ for some polynomial q over F. If R contains 0, the polynomial with all coefficients zero, we set $r = 0 = f - qg$, and we are finished except for uniqueness.

Suppose then $0 \notin R$. Then the set

$$S = \{n \in \tilde{\mathbf{N}} \mid n = \deg h, h \in R\}$$

is nonempty since either $\deg f \in S$ or, when $f = 0$, $\deg g \in S$. Therefore, S has a smallest element m. By definition of S we have $m = \deg r$ for some $r \in R$, and by definition of R we have $r = f - qg$ for some q. In other words, $f = qg + r$, and it remains to show that $\deg r < \deg g$. Suppose that $m = \deg r \ge \deg g = n$. Clearly, there is an element $c \in F$ such that the polynomial s given by

$$sx = rx - cx^{m-n}(gx) = fx - [(qx) + (cx^{m-n})](gx)$$

has degree $m - 1$ or less. However, $s \in R$ and this contradicts the minimality of m. It must be that deg $r <$ deg g.

Suppose q' and r' are polynomials satisfying the same conditions as q and r. Then $f = qg + r = q'g + r'$ implies

$$(q - q')g = r' - r.$$

If $q - q' \neq 0$, then taking degrees on both sides of this last equation we must have

$$\deg(q - q') + \deg g = \max\{\deg r', \deg r\}.$$

This implies that either deg $g \leq$ deg r' or deg $g \leq$ deg r, both of which are wrong. Thus, we must have $q = q'$ and, consequently, $r = r'$. This proves uniqueness.

In practice it is not difficult to determine q and r: we simply carry out the customary long division of f by g obtaining q as the quotient and r as the remainder. If $r = 0$, then we say that g *divides* f and we write $g \,|\, f$.

Corollary. (**The Remainder Theorem**) *If f is a polynomial over the field F and α is an element of F, then there is a unique polynomial q over F such that*

$$fx = (x - \alpha)(qx) + (f\alpha).$$

Proof. Applying the division theorem with g given by $gx = x - \alpha$, we have

$$fx = (x - \alpha)(qx) + (rx)$$

where $r = 0$ or deg $r <$ deg $g = 1$. Thus, rx is a constant, and taking $x = \alpha$ shows that $rx = f\alpha$.

99α. Indicate the changes needed in the proof of the theorem above to prove the following: *if f and g are polynomials with integer coefficients and g is monic, then there are unique polynomials q and r with integer coefficients such that $f = qg + r$ where either $r = 0$ or* deg $r <$ deg g. (A polynomial is *monic* if the leading coefficient is 1.)

100. An element α of the field F is a *root* of the polynomial f over F if $f\alpha = 0$. In other words, α is a root of f if f, considered as a function, assigns the value 0 to α.

The remainder theorem just proved implies: *if α is a root of f, then $fx = (x - \alpha)(qx)$, or in other words, $(x - \alpha)$ divides f.*

Proposition.   *A polynomial of degree n over the field F has at most n roots in F.*

Proof.   The proof is by induction on n. To start the induction, we note that a polynomial of degree 0 is a (nonzero) constant and has no roots. Now suppose the proposition is true for polynomials of degree less than n. Let f be a polynomial of degree n. If f has no roots in F, then we are finished. If f has a root $\alpha \in F$, then $fx = (x - \alpha)(qx)$ for some polynomial q over F of degree $n - 1$. Then q has at most $n - 1$ roots in F; it clearly follows that f can have at most n roots since a root of f is either α or a root of q.

A polynomial of degree n over the field F which has all n roots in F is said to *split* over F. Clearly, a polynomial f of degree n splits over F if and only if it can be factored as

$$fx = c(x - \alpha_1)(x - \alpha_2) \cdots (x - \alpha_n)$$

where $\alpha_1, \alpha_2, \ldots, \alpha_n \in F$ are the roots of f.

Theorem.   *The multiplicative group F^* of a finite field F is cyclic.*

Proof.   Since F^* is abelian, each of its Sylow subgroups is normal, and therefore, for a prime dividing the order of F^*, there is just one Sylow subgroup. It follows from **59θ** that F^* is the direct product of its Sylow subgroups. Furthermore, the orders of the Sylow subgroups are relatively prime, and therefore, it follows from **43γ** that F^* is cyclic if each Sylow subgroup is. To see this, let H denote the p-Sylow subgroup for a prime p dividing the order of F^*, and let α be an element of maximal order, say p^k, in H. Then the order of every element of H must divide p^k. From the preceding proposition it follows that there cannot be more than p^k elements of F satisfying the equation $x^{p^k} = 1$, and thus $o(H) = p^k = o(\alpha)$, from which we conclude that H is cyclic.

Exercise **100ε** suggests an alternate proof of this theorem.

100α.   The *formal derivative* of the polynomial $fx = c_0 + c_1 x + \cdots + c_n x^n$ is the polynomial $f'x = c_1 + 2c_2 x + \cdots + nc_n x^{n-1}$. Verify the rules of formal differentiation: $(f + g)' = f' + g'$ and $(fg)' = f'g + fg'$.

100β.   Show that a polynomial f over a field F and its derivative f' have a common root α in F if and only if α is a *multiple root* of f, that is, $(x - \alpha)^2$ divides f.

100γ.   Show that there exists one and only one polynomial of degree n or less over a field F which assumes $n + 1$ prescribed values $f\alpha_0 = \beta_0$, $f\alpha_1 = \beta_1$, ..., $f\alpha_n = \beta_n$ where $\alpha_0, \alpha_1, \ldots, \alpha_n$ are distinct elements of F. (The expression for f is called the *Lagrange interpolation formula.*)

100δ.   Show that every element of a finite field with q elements is a root of the polynomial $x^q - x$.

100ε. By counting the number of elements of F which are roots of $x^n - 1$ for various values of n, show that *the multiplicative group F^* of a finite field F is cyclic.* (See **25α**.)

100ζ. For p a prime construct a group isomorphism $\mathbf{Z}_{p-1} \to \mathbf{Z}'_p$.

100η. For p a prime show that $(p - 1)! \equiv -1 \bmod p$. (This is known as *Wilson's theorem* after Sir John Wilson (1741–1793), who was a student of Edward Waring (1736–1798). The statement, but not the proof, of this theorem first appeared in Waring's *Meditationes Algebraicae* of 1770 (p. 218). The first published proof is due to Lagrange in 1771. Lagrange also proved the converse: if $(n - 1)! \equiv -1 \bmod n$, then n is prime.)

100θ. Show that a polynomial $fx = c_0 + c_1 x + \cdots + c_n x^n$ over $\mathbf{Z}$ has a root $p/q \in \mathbf{Q}$, where $p, q \in \mathbf{Z}$ and $(p, q) = 1$, only if $p \mid c_0$ and $q \mid c_n$.

100ι. Let f be a polynomial over a field F whose derivative (**100α**) is 0. Show that if char $F = 0$, then f is a constant polynomial. What can one say in the case where char $F \neq 0$?

101. In general a polynomial of degree n over a field may have any number of roots from 0 to n in that field. A notable exception to this occurs for the field of complex numbers $\mathbf{C}$.

The Fundamental Theorem of Algebra. *A polynomial of positive degree over the field $\mathbf{C}$ of complex numbers has a root in $\mathbf{C}$.*

Proof. Unfortunately, all proofs of this theorem use analysis and therefore are not really algebraic. We shall give a proof due to Ankeny which uses the theorem of Cauchy from complex function theory. (This proof is included for completeness only and may be skipped by the reader unfamiliar with complex function theory.)

Let f be a polynomial of degree $n \geq 1$ over $\mathbf{C}$ given by

$$fz = c_0 + c_1 z + \cdots + c_n z^n.$$

We let $\bar{f}$ denote the polynomial of degree n over $\mathbf{C}$ whose coefficients are the complex conjugates of those of f; that is,

$$\bar{f}z = \bar{c}_0 + \bar{c}_1 z + \cdots + \bar{c}_n z^n.$$

Now the product $\phi = f\bar{f}$ is a polynomial of degree $2n$ over $\mathbf{C}$ with *real* coefficients. (Why?) We observe that *it is sufficient to prove that ϕ has a root*: if $\phi\alpha = (f\alpha)(\bar{f}\alpha) = 0$, then either $f\alpha = 0$ (and f has α as a root) or $\bar{f}\alpha = 0$ and $f\bar{\alpha} = 0$.

Suppose ϕ has no root in **C**. Then the complex function $1/\phi$ is analytic in the whole complex plane. It follows from Cauchy's theorem that the integral of $1/\phi$ along any path in the plane depends only upon the endpoints of the path. In particular, the integral of $1/\phi$ around the upper half (Γ) of the circle $|z| = R$ in a clockwise direction equals the integral of $1/\phi$ along the real axis from $-R$ to R.

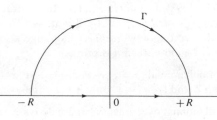

Figure 5

We examine the behavior of these integrals as R grows large. Since $\deg \phi = 2n$, we can write $\phi(z) = az^{2n} - \psi(z)$ where $a \neq 0$ and ψ is a polynomial of degree less than $2n$. Consequently, we have

$$\left| \frac{\phi(z)}{az^{2n}} \right| \geq 1 - \left| \frac{\psi(z)}{az^{2n}} \right|.$$

Suppose now that $\psi(z) = a_0 + a_1 z + \cdots + a_m z^m$ where $m < 2n$. Then

$$\left| \frac{\psi(z)}{az^{2n}} \right| \leq \frac{|a_0| + |a_1| \, |z| + \cdots + |a_m| \, |z|^m}{|a| \, |z|^{2n}}$$

$$\leq \frac{(|a_0| + |a_1| + \cdots + |a_m|)}{|a| \, |z|^{2n-m}}$$

at least when $|z| > 1$. It follows that for any $\varepsilon \in (0,1)$, there exists some $R_\varepsilon > 1$ such that $|z| \geq R_\varepsilon$ implies $|\psi(z)/az^{2n}| \leq \varepsilon$, and also

$$|\phi(z)| \geq |az^{2n}|(1 - \varepsilon) \geq |a|R_\varepsilon^{2n}(1 - \varepsilon).$$

We apply this to $\int_\Gamma dz/\phi(z)$ for $R \geq R_\varepsilon$ to get

$$\left| \int_\Gamma \frac{dz}{\phi(z)} \right| \leq \int_\Gamma \frac{|dz|}{|\phi(z)|} \leq \int_\Gamma \frac{|dz|}{|a|R^{2n}(1 - \varepsilon)} = \frac{\pi}{|a|R^{2n-1}(1 - \varepsilon)}.$$

Thus, as R grows large, $\int_\Gamma dz/\phi(z)$ grows small in absolute value.

Now consider the other integral. Since ϕ has real coefficients, it takes only real values along the real axis. Furthermore, ϕ cannot change sign along the real axis—to do so it would have to vanish somewhere, contrary to the hypothesis that ϕ has no roots. It follows that the integral of $1/\phi$ along the x axis, which can be expressed as

$$\int_{-R}^{+R} \frac{dx}{\phi(x)},$$

can only increase in absolute value as R grows. Of course this behavior is completely opposite to that of the integral around Γ and provides the contradiction which establishes the theorem.

Corollary. *A polynomial f of degree n over* **C** *has n roots in* **C** *and factors as*

$$fx = c(x - \alpha_1)(x - \alpha_2) \cdots (x - \alpha_n),$$

where $\alpha_1, \alpha_2, \ldots, \alpha_n$ are the roots of f and $c \in$ **C***.*

Proof. By the theorem, f has a root $\alpha_1 \in$ **C**. Then $fx = (x - \alpha_1)(gx)$ where g is a polynomial of degree $n - 1$. Again, g has a root $\alpha_2 \in$ **C** and $gx = (x - \alpha_2)(hx)$, and so forth.

N.B. *The roots $\alpha_1, \alpha_2, \ldots, \alpha_n$ need not be distinct.* A number which occurs more than once in the list of roots is called a *repeated* or *multiple root*. Those which occur once are called *simple roots*.

The fundamental theorem of algebra was stated first in 1746 by the Frenchman Jean-le-Rond D'Alembert (1717–1783), who gave an incomplete proof. The first true proof was given by Karl Friedrich Gauss (1777–1855) in 1799. Gauss gave, in all, four distinct proofs of this theorem.

A field with the property that every polynomial over it splits into linear factors is said to be *algebraically closed*. The fundamental theorem may therefore be restated as: *the field of complex numbers is algebraically closed*. It is true, but difficult to prove, that every field is contained in an algebraically closed field.

101α. Show that every polynomial over **R** of positive degree can be factored as a product of polynomials over **R** with degree 1 or 2.

101β. A number α is a *root of multiplicity m* of a polynomial ϕ over **C** if

$$(x - \alpha)^m | \phi x \quad \text{but} \quad (x - \alpha)^{m+1} \nmid \phi x.$$

Show that α is a root of ϕ of multiplicity m if and only if $\phi\alpha = \phi'\alpha = \cdots = \phi^{(m-1)}\alpha = 0$, but $\phi^{(m)}\alpha \neq 0$. (Here $\phi^{(k)}$ denotes the k-th derivative of ϕ.)

102.   A polynomial g over the field F *divides* a polynomial f over F if $f = qg$ for some polynomial q over F. To indicate that g divides f, we write $g \mid f$, and to indicate that it does not, $g \nmid f$. A polynomial is always divisible by itself and by every polynomial of degree 0.

A polynomial f over F of positive degree which can be factored as $f = gh$ where g and h are polynomials over F of *positive* degree is called *reducible over F*; a polynomial of positive degree which cannot be thus factored is called *irreducible over F*. (We shall not apply either term to polynomials of degree zero.) Every polynomial of degree 1 is irreducible. In general there are many irreducible polynomials of higher degrees over a field. As we shall see, irreducible polynomials are like prime numbers.

Proposition.   *A polynomial f, irreducible over the field F, has a root in F if and only if* $\deg f = 1$.

Proof.   If $\deg f = 1$, then $fx = c_0 + c_1 x$ and f has $-c_0/c_1$ as root in F. On the other hand, if f has a root $\alpha \in F$, then $fx = (x - \alpha)(qx)$ for some polynomial q over F. Since f is irreducible, it must be that $\deg q = 0$ and, consequently, $\deg f = 1$.

Corollary.   *The only irreducible polynomials over the field of complex numbers* **C** *are those of degree* 1.

As an example we note that the polynomial $x^2 + 1$ over the field of rational numbers **Q** is irreducible, but considered as a polynomial over **C** it is reducible: $x^2 + 1 = (x - i)(x + i)$.

102α.   Show that a polynomial irreducible over **R** has degree 1 or 2.

102β.   Show that every polynomial of positive degree over a field F is divisible by a polynomial irreducible over F.

102γ.   Show that there are an infinite number of irreducible polynomials over any field.

102δ.   Compute the number of irreducible polynomials of degrees 1, 2, and 3 over $\mathbf{Z}_p$.

102ε.   Determine all of the *monic* polynomials (that is, polynomials with leading coefficient 1) of degrees 2 and 3 which are irreducible over $\mathbf{Z}_3$.

102ζ.   Show that the polynomial f over a field is irreducible if and only if the polynomial g defined by $gx = f(x + a)$ is irreducible over the same field.

102η.   Show that $4x^3 - 3x - 1/2$ is irreducible over **Q**.

102θ.   An integer m is called a *quadratic residue* mod p if and only if the congruence $x^2 \equiv m$ mod p has a solution, or what is the same thing, if and only if the polynomial $x^2 - [m]_p$ has a root in the field Z_p. Count the number of elements $[m]_p \in Z_p$ for which $x^2 - [m]_p$ has a root in Z_p.

102ι. Show that $x^2 - a$ has a root in Z_p ($p > 2$) if and only if $a^{(p+1)/2} = a$.

102κ. Determine conditions on a and b for which the quadratic equation $x^2 + ax + b = 0$ is solvable in Z_p.

103. A *greatest common divisor* of two polynomials of positive degree over the field F is a polynomial of maximal degree over F dividing both. That is, d is a greatest common divisor of f and g if $d | f$ and $d | g$, but $\deg h > \deg d$ implies either $h \nmid f$ or $h \nmid g$. For a rather trivial reason, there is more than one polynomial which satisfies these requirements: *if d is a greatest common divisor of f and g over F and $c \in F^*$, then the polynomial cd (given by $(cd)x - c(dx)$) is also a greatest common divisor.* We shall let (f, g) denote the *set* of polynomials which are greatest common divisors of f and g over F. Outside of this aspect of the situation, the notion of greatest common divisor for polynomials is similar to that for integers (**23**).

Theorem. *If f and g are polynomials of positive degree over F and $d \in (f, g)$, then there exist polynomials u and v over F such that*

$$d = uf + vg.$$

Proof. Let $\mathfrak{a}$ denote the set of polynomials of the form $sf + tg$ over F. Let

$$\mathfrak{b} = \{n \in \tilde{N} \mid n = \deg h, h \in \mathfrak{a}, h \neq 0\}.$$

The set $\mathfrak{b}$ contains $\deg f$ and $\deg g$ and therefore has a smallest element $m = \deg d'$, where $d' = u'f + v'g \in \mathfrak{a}$ for some polynomials u' and v' over F. We must have $d' | h$ for all $h \in \mathfrak{a}$. Otherwise, for some h, $h = qd' + r$ where $\deg r < \deg d'$, and if $h = sf + tg$, we have

$$r = h - qd' = (s - qu')f + (t - qv')g$$

so that $r \in \mathfrak{a}$, contradicting the minimality of $\deg d'$ in $\mathfrak{b}$. Thus, d' divides every element of $\mathfrak{a}$—in particular $d' | f$ and $d' | g$. Therefore, $\deg d' \leq \deg d$. On the other hand, $d | f$ and $d | g$ so that d divides $d' = u'f + v'g$ and $\deg d \leq \deg d'$. Consequently, $\deg d = \deg d'$. However, $d | d'$ implies $d' = cd$ where $c \in F$. Setting $u = u'/c$ and $v = v'/c$, we have $d = d'/c = uf + vg$.

Corollary. *If $d, d' \in (f, g)$, then $d' = cd$ for some nonzero $c \in F$.*

Corollary. *If f, g, and h are polynomials over the field F, if f is irreducible over F, and if f divides gh, then f divides g, or f divides h (or both).*

106. **Proposition.** *Every nonzero polynomial f over the field $\mathbf{Q}$ of rational numbers can be written uniquely as $f = c\bar{f}$ where $c \in \mathbf{Q}$, $c > 0$, and $\bar{f}$ is a primitive polynomial.*

The positive rational number c is called the *content of f*, and the polynomial $\bar{f}$ is called the *primitive form* of f.

Proof. Clearly, $f = af'$ where $a \in \mathbf{Q}$, $a > 0$, and f' is a polynomial with integral coefficients. Let $\bar{f}$ be the polynomial obtained from f' by dividing each coefficient by the number $b \in \mathbf{N}$, which is the greatest common divisor of them all. Then $f = c\bar{f}$ where $c = ab$.

Suppose f can be written in two ways: $f = c\bar{f} = d\bar{g}$ where $\bar{f}$ and $\bar{g}$ are primitive. Let $c = p/q$ and $d = r/s$ where p, q, r, $s \in \mathbf{N}$. Then $sp\bar{f} = qr\bar{g}$ is a polynomial with integral coefficients having greatest common divisor $sp = qr$. It follows that $c = p/q = r/s = d$ and, consequently, that $\bar{f} = \bar{g}$.

Corollary. *A nonzero polynomial with integral coefficients is reducible over $\mathbf{Q}$ if and only if it factors as a product of two polynomials with integral coefficients of positive degree.*

Proof. Suppose f is a polynomial over $\mathbf{Z}$ which is reducible over $\mathbf{Q}$, say $f = gh$. Let $g = a\bar{g}$ and $h = b\bar{h}$ be the factorizations of g and h guaranteed by the proposition. Then $f = (ab)\bar{g}\bar{h}$ is a factorization of f since $\bar{g}\bar{h}$ is primitive. Therefore, ab is the content of f and therefore an integer. We have $f = ((ab)\bar{g})\bar{h}$, the required factorization. The argument in the other direction is trivial.

This corollary is called *Gauss's lemma* because it is given in article 42 of his famous *Disquisitiones Arithmeticae* of 1801. The proposition of **105** is also called Gauss's lemma by some authors.

107. **The Eisenstein Irreducibility Criterion.** *Let f be a polynomial over $\mathbf{Q}$ with integral coefficients, say $fx = c_0 + c_1 x + \cdots + c_n x^n$. If there is a prime number p such that p divides every coefficient of f except c_n and p^2 does not divide c_0, then f is irreducible over $\mathbf{Q}$.*

Proof. Suppose that f is reducible. Then by the corollary of **106**, f must factor as $f = gh$ where g and h are polynomials of positive degree with integer coefficients. Let g and h be given by

$$gx = a_0 + a_1 x + \cdots + a_r x^r$$

and

$$hx = b_0 + b_1 x + \cdots + b_s x^s, \qquad r + s = n.$$

The coefficients are related by the equations

$$c_0 = a_0 b_0,$$
$$c_1 = a_0 b_1 + a_1 b_0,$$
$$\vdots \qquad \vdots$$
$$c_n = a_0 b_n + a_1 b_{n-1} + \cdots + a_n b_0.$$

By hypothesis, $p \mid c_0$ and therefore $p \mid a_0$ or $p \mid b_0$, but not both since $p^2 \nmid c_0$. Without loss of generality, we may assume that $p \mid a_0$ and $p \nmid b_0$. Now $p \mid c_1$ and $p \mid a_0$ imply $p \mid a_1 b_0$; since $p \nmid b_0$, it follows that $p \mid a_1$. Continuing in this fashion, we obtain $p \mid a_0, p \mid a_1, \ldots, p \mid a_r$. Thus, p divides every coefficient of g. Since $f = gh$, it follows that p divides every coefficient of f. But this contradicts the hypothesis that $p \nmid c_n$. Thus, f cannot be reducible, and the proof is complete.

Ferdinand Gotthold Max Eisenstein (1823–1852) was a student of Gauss and continued the work of the master begun in the *Disquisitiones Arithmeticae*. The theorem above appeared in Crelle's *Journal für Mathematik* vol. 39 (1850), pp. 160–179. It is sometimes erroneously attributed to Theodor Schönemann (1812–1868).

107α. Prove that there exist a countable number of irreducible polynomials of degree n over $\mathbf{Q}$.

107β. Show that the polynomial $\Phi_p x = 1 + x + \cdots + x^{p-1}$ is irreducible over $\mathbf{Q}$ for p a prime. (*Hint*: consider $gx = \Phi_p(x + 1)$.)

107γ. By means of the Eisenstein criterion, show that the cubic $4x^3 - 3x - 1/2$ is irreducible over $\mathbf{Q}$.

107δ. Show that a polynomial of odd degree $2m + 1$ over $\mathbf{Z}$,

$$fx = c_0 + c_1 x + \cdots + c_{2m+1} x^{2m+1},$$

is irreducible if there exists a prime p such that

(1) $p \nmid c_{2m+1}$,
(2) $p \mid c_{m+1}, p \mid c_{m+2}, \ldots, p \mid c_{2m}$,
(3) $p^2 \mid c_0, p^2 \mid c_1, \ldots, p^2 \mid c_m$,
(4) $p^3 \nmid c_0$.

(This is a theorem of Eugen Netto (1846–1919) and appears in *Mathematische Annalen*, vol. 48 (1897).)

107ε. Let

$$\phi X = (f_0 x) + (f_1 x)X + \cdots + (f_n x)X^n$$

be a polynomial in X over the field $F(x)$ (**98α**) with coefficients in $F[x]$. Suppose that x divides $f_0 x, f_1 x, \ldots, f_{n-1} x$ but not $f_n x$, and that x^2 does not divide $f_0 x$. Prove that ϕX is irreducible over $F(x)$.

Algebraic Extensions

---

108.   Let E be an extension field of the field F. An element α of E is *algebraic over F* if α is a root of some polynomial with coefficients in F. If every element of E is algebraic over F, then E is called an *algebraic extension* of F.

As examples we note that $\sqrt{2}$ and $i = \sqrt{-1}$ are algebraic over $\mathbf{Q}$. Complex numbers which are algebraic over the rational field $\mathbf{Q}$ are called *algebraic numbers*. There exist complex numbers which are not algebraic (e and π for example) and these are called *transcendental numbers*.

108α.   Prove that the sum, $c + \alpha$, and product, $c\alpha$, of a rational number c and an algebraic number α are algebraic numbers.

108β.   Prove that $\cos(k\pi)$ is an algebraic number whenever k is rational.

109.   Let α be an element of the extension field E of the field F, and suppose α is algebraic over F. Among all the polynomials over F of which α is a root, let f be one of lowest degree. Then f is called a *minimal polynomial* for α over F. Minimal polynomials have two important properties.

Proposition.   *If f is a minimal polynomial for α over F, then*

(1)   *f is irreducible over F,*
(2)   *f divides any polynomial over F having α as a root.*

Proof.   Suppose f is reducible, say $f = gh$. Then we have $f\alpha = (g\alpha)(h\alpha) = 0$, which implies $g\alpha = 0$ or $h\alpha = 0$. Both g and h have degree less than f, contradicting the definition of f as a minimal polynomial for α. Thus, f is irreducible.

Suppose α is a root of a polynomial g over F. By the division theorem we can write $g = qf + r$. Then we have $g\alpha = (q\alpha)(f\alpha) + r\alpha = 0$ which implies $r\alpha = 0$. If $r \neq 0$, then $\deg r < \deg f$. But then r is a polynomial of degree less than f with α as a root, contradicting the minimality of f. Thus, $r = 0$ and $f \mid g$.

Corollary.   *Two minimal polynomials for α over F differ by a constant factor.*

109α.   Let f be a polynomial irreducible over F, and let E be an extension field of F in which f has a root α. Show that f is a minimal polynomial for α over F.

109β.   Let $F \subset E \subset D$ be a tower of fields. Let $\alpha \in D$, and let g be a minimal polynomial for α over E and f a minimal polynomial for α over F. Show that $g \mid f$ (considering both as polynomials over E).

109γ. Find minimal polynomials over **Q** and $\mathbf{Q}(\sqrt{2})$ for the numbers $\sqrt{2} + \sqrt{3}$ and $i\sqrt{2} = \sqrt{-2}$.

110. Let α be an element of E, an extension field of F. We denote by $F(\alpha)$ the smallest subfield of E containing both F and α. $F(\alpha)$ is called *the field obtained by adjoining α to F*. We may also characterize $F(\alpha)$ as the intersection of all the subfields of E which contain α and F.

Proposition. *If E is an extension field of F and $\alpha \in E$ is algebraic over F, then $F(\alpha)$ is a finite extension of F of degree n where n is the degree of a minimal polynomial for α over F. Furthermore, the set $\{1, \alpha, \alpha^2, \ldots, \alpha^{n-1}\}$ is a basis for $F(\alpha)$ over F.*

Proof. Since $F(\alpha)$ is a field and contains α, it must contain all the elements $1, \alpha, \alpha^2, \ldots, \alpha^{n-1}$ and therefore, as a vector space over F, it must contain every linear combination

$$c_0 + c_1\alpha + \cdots + c_{n-1}\alpha^{n-1},$$

with coefficients in F. Let X denote the set of all such linear combinations. It is not difficult to see that X is a vector space over F spanned by

$$\{1, \alpha, \ldots, \alpha^{n-1}\}.$$

Now we assert that $\{1, \alpha, \ldots, \alpha^{n-1}\}$ is linearly independent over F. If there were a nontrivial linear relation over F,

$$c_0 + c_1\alpha + \cdots + c_{n-1}\alpha^{n-1} = 0,$$

then α would be a root of the polynomial g over F given by

$$gx = c_0 + c_1 x + \cdots + c_{n-1}x^{n-1}.$$

However, $\deg g < n$, and by hypothesis n is the degree of a minimal polynomial for α over F. This contradiction forces the conclusion that

$$\{1, \alpha, \alpha^2, \ldots, \alpha^{n-1}\}$$

is linearly independent and hence that it is a basis for X over F.

The remainder of the proof consists of showing that X is a field. Since X contains F and α, this implies that $F(\alpha) \subset X$. We already know that $X \subset F(\alpha)$, so that we will have $X = F(\alpha)$.

Clearly, X is an additive subgroup of E. To show that X is a subfield of E, and hence a field, we need only verify that $X^* = X - \{0\}$ is a multiplicative subgroup of $E^* = E - \{0\}$. Let f be a minimal polynomial for α over F. Suppose that

$$\beta = b_0 + b_1\alpha + \cdots + b_{n-1}\alpha^{n-1}$$

and

$$\gamma = c_0 + c_1\alpha + \cdots + c_{n-1}\alpha^{n-1}$$

are elements of X^*. We can write $\beta = g\alpha$ and $\gamma = h\alpha$ for the polynomials g and h over F given by

$$gx = b_0 + b_1x + \cdots + b_{n-1}x^{n-1}$$

and

$$hx = c_0 + c_1x + \cdots + c_{n-1}x^{n-1}.$$

By the division theorem we have $gh = qf + r$ where $r = 0$ or $\deg r < \deg f = n$. Since $f\alpha = 0$, we have

$$0 \neq \beta\gamma = (g\alpha)(h\alpha) = (gh)\alpha = (qf)\alpha + r\alpha = r\alpha.$$

Since $r\alpha \neq 0$, we have $r \neq 0$, and consequently, $\deg r < n$. Thus,

$$\beta\gamma = r\alpha = a_0 + a_1\alpha + \cdots + a_{n-1}\alpha^{n-1} \in X^*.$$

Finally, we show that every element of X^* has a multiplicative inverse. Let

$$\beta = g\alpha = b_0 + b_1\alpha + \cdots + b_{n-1}\alpha^{n-1}$$

as above. By **109**, the minimal polynomial f for α over F is irreducible. Therefore, $1 \in (f, g)$ the greatest common divisor, and by **103** there exist polynomials u and v over F such that $uf + vg = 1$. Moreover, we can find u and v so that $\deg v < \deg f = n$. Since $f\alpha = 0$, we obtain $(v\alpha)(g\alpha) = 1$. Thus, $\beta^{-1} = v\alpha \in X^*$.

As an example of this proposition, consider the field $\mathbf{Q}(\zeta)$ where $\zeta = e^{2\pi i/p}$ for p prime. Now ζ is a p-th root of unity, that is, a root of $x^p - 1$, and therefore is algebraic over $\mathbf{Q}$. We have the factorization

$$x^p - 1 = (x - 1)\Phi_p(x)$$

where

$$\Phi_p(x) = x^{p-1} + x^{p-2} + \cdots + 1.$$

It follows that ζ is a root of Φ_p, which is irreducible over $\mathbf{Q}$ as we shall see. The substitution $x = y + 1$ yields

$$\Phi_p(y+1) = \frac{(y+1)^p - 1}{(y+1) - 1}$$

$$= y^{p-1} + \binom{p}{1} y^{p-2} + \cdots + \binom{p}{p-2} y + \binom{p}{p-1},$$

to which the Eisenstein criterion applies using the prime p. Since Φ_p is irreducible, it must be the minimal polynomial for ζ over $\mathbf{Q}$. Consequently, $[\mathbf{Q}(\zeta): \mathbf{Q}] = p - 1$ and $\{1, \zeta, \zeta^2, \ldots, \zeta^{p-2}\}$ is a basis for $\mathbf{Q}(\zeta)$ over $\mathbf{Q}$.

110α. Let E be an extension field of F, and let $\alpha \in E$ be an element algebraic over F. Show that $F(\alpha)$ is isomorphic to the field $F[x]/(f)$ where f is a minimal polynomial of α over F. (See **103β**.)

110β. Let $\alpha, \beta \in E$ be elements algebraic over the subfield F. It is clear that β is algebraic over $F(\alpha)$. We denote by $F(\alpha, \beta)$ the subfield of E obtained by adjoining β to $F(\alpha)$. Show that $F(\alpha, \beta) = F(\beta, \alpha)$. What can be said of the degree $[F(\alpha, \beta): F]$?

110γ. Let E be an extension field of F which contains all the roots, $\alpha_1, \alpha_2, \ldots, \alpha_n$, of a polynomial f of degree n. The *splitting field* of f in E is the smallest subfield of E containing F and the roots $\alpha_1, \alpha_2, \ldots, \alpha_n$. We denote this by $F(\alpha_1, \alpha_2, \ldots, \alpha_n)$. Prove that

$$[F(\alpha_1, \alpha_2, \ldots, \alpha_n): F] \leq n!.$$

110δ. Let E and E' be two extensions of F in which a polynomial f over F splits. Prove that there exists an isomorphism ϕ from the splitting field of f in E to the splitting field of f in E', such that $\phi c = c$ for every $c \in F$.

110ε. Let $\alpha \in E$ be an element *transcendental* (that is, *not* algebraic) over a subfield F. Prove that $F(\alpha)$ is a field isomorphic to $F(x)$, the field of rational functions of x over F (**98α**).

110ζ. Prove that two finite fields with the same number of elements are isomorphic.

110η. Let $\alpha \in E$ be an element transcendental over the subfield F. What is the degree of $F(\alpha)$ over $F(\alpha^4/4\alpha^3 - 1)$?

111. Proposition. *A finite extension is an algebraic extension.*

Proof. Let E be a finite extension of the field F, and suppose that $[E: F] = n$. Let $\alpha \in E$ be any element. The set of $n + 1$ elements

$$\{1, \alpha, \alpha^2, \ldots, \alpha^n\}$$

(0) *The points* $(0, 0)$ *and* $(1, 0)$ *are constructible.* (Any two points of the plane may be chosen for $(0, 0)$ and $(1, 0)$ and the distance between them taken as the unit length.)

(1) *The line (or line segment) determined by two constructible points is constructible.*

(2) *A circle with a constructible point as center and a constructible length as radius is constructible.* (A constructible length is the distance between two constructible points.)

(3) *The intersection of two constructible lines is a constructible point.*

(4) *The points (or point) of intersection of a constructible line and a constructible circle are constructible.*

(5) *The points (or point) of intersection of two constructible circles are constructible.*

Remarks. We shall call (0)–(5) the *axioms of constructibility.* Once they have been stated, the problem of constructibility with straightedge and compass is removed from the domain of mechanical drawing to the domain of mathematics. Axiom 1 indicates the only way the straightedge may be used: to draw the line between two previously constructed points. Axiom 2 indicates how the compass is used: the feet may be placed on two constructed points to determine a radius and then the compass transported to a third constructed point as center and the circle drawn. Axioms 3, 4, and 5 indicate the ways in which new points are constructed. A warning to the reader may prevent misinterpretation: *lines and circles are not to be considered as "made up" of points; that a line or circle is constructible does not imply that all points on the line or circle are constructible. Furthermore, we do not allow the choice of arbitrary points on or off lines or circles.*

116. Proposition. *The line parallel to a given constructible line and passing*

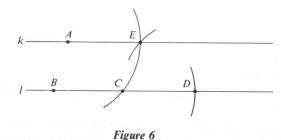

Figure 6

through a given constructible point (not on the given line) is constructible.

Proof. Let A be a constructible point and l a constructible line not passing through A. Let B and C be constructible points which determine l. The circle with center A and radius AC is constructible. The circle with center C and radius AC is also constructible and so are its intersections with the line l. Let D be one of these intersections. The circle with center D and radius $CD = AC$ is constructible and intersects the circle with center A and radius AC in the points C and E. Thus, the point E is constructible. Finally, the line k determined by A and E is constructible and is parallel to l.

117. Proposition. *The perpendicular bisector of a constructible line segment*

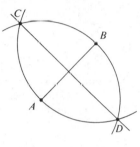

Figure 7

is a constructible line.

Proof. Let A and B be constructible points. The circles centered at A and B with radius AB are constructible and so are their intersection points C and D. The line determined by C and D is constructible and is the perpendicular bisector of AB.

118. Proposition. *The circle determined by three constructible points (not*

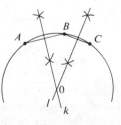

Figure 8

lying in a line) is constructible.

Proof.   Let A, B, and C be three constructible points which do not lie on a line. By **117**, the perpendicular bisectors k and l of the line segments AB and AC are constructible. Consequently, their point of intersection is a constructible point O. The circle with center O and radius $AO = BO = CO$ is constructible and passes through A, B, and C.

119.   We shall call a complex number $a + bi$ *constructible*, if the corresponding coordinate point (a, b) is constructible with straightedge and compass according to the axioms of **115**. The complex numbers 0 and 1 are constructible by axiom 0.

Theorem.   *The constructible numbers form a field* $\mathscr{C}$.

Proof.   Since the numbers 0 and 1 are constructible, the real axis (which they determine) is a constructible line. Clearly, the number -1 is constructible. The perpendicular bisector of the segment between -1 and 1 is constructible, so that the axis of imaginary numbers is a constructible line.

First, we show that the real numbers which are constructible form a field. It is obvious that if a and b are constructible real numbers, then $a + b$ and $-a$ are constructible. Suppose that a and b are constructible, positive real numbers. Then the numbers $ai = (0, a)$ and $-bi = (0, -b)$ are constructible. (Why?) By **118** the circle through $(-1, 0)$, $(0, a)$, and $(0, -b)$ is constructible.

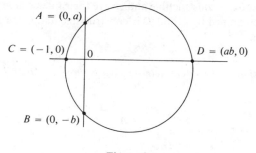

Figure 9

This circle intersects the real axis in a constructible point D. AB and CD are chords of the circle intersecting at the origin O. By a theorem of elementary geometry $(AO)(OB) = (CO)(OD)$. It follows that $OD = ab$ and $D = (ab, 0)$. Thus, ab is constructible. In a similar manner, $1/a$ is constructible as shown in Figure 10.

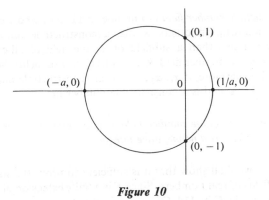

Figure 10

This completes showing that the constructible real numbers form a field.

It is clear that the complex number $a + bi$ is constructible if and only if the real numbers a and b are constructible. (We need to use **116** here.) It follows

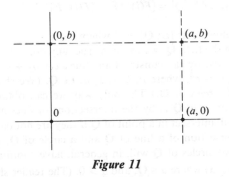

Figure 11

immediately from this observation that when $a + bi$ and $c + di$ are constructible complex numbers, the numbers

$$(a + bi) + (c + di) = (a + c) + (b + d)i,$$
$$-(a + bi) = (-a) + (-b)i$$
$$(a + bi)(c + di) = (ac - bd) + (ad + bc)i,$$

and

$$\frac{1}{a + bi} = \left(\frac{a}{a^2 + b^2}\right) = \left(\frac{b}{a^2 + b^2}\right)i$$

are also constructible.

To demonstrate the general impossibility of trisecting angles with straight-edge and compass, it is sufficient to exhibit one angle for which trisection is impossible. We choose 60°.

First, we observe that an angle α is constructible if and only if the real number $\cos \alpha$ is constructible. Thus, we have only to show that $\cos 20°$ is not

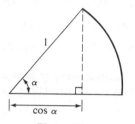

Figure 13

constructible. For any angle θ, we have $\cos 3\theta = 4 \cos^3 \theta - 3 \cos \theta$. Since $\cos 60° = 1/2$, it happens that $\cos 20°$ is a solution of the equation $4x^3 - 3x = 1/2$, or what is the same thing, a root of the polynomial $8x^3 - 6x - 1$.

The polynomial $8x^3 - 6x - 1$ is irreducible over **Q**: the substitution $x = \frac{1}{2}(y + 1)$ yields $y^3 - 3y^2 - 3$, which is clearly irreducible by the Eisenstein criterion (**107**); were $8x^3 - 6x + 1$ reducible over **Q**, the same applied to its factors would yield a factorization of $y^3 - 3y^2 - 3$.

Now it follows that $8x^3 - 6x - 1$ is a minimal polynomial for $\cos 20°$ over **Q** and that $[\mathbf{Q}(\cos 20°) : \mathbf{Q}] = 3$. By **120** we are forced to conclude that **Q**($\cos 20°$) is not a constructible number field and that $\cos 20°$ is not a constructible number. Thus, the angle 20° cannot be constructed.

There are four famous problems of antiquity concerned with straightedge and compass constructions. One is the trisectability of angles, which we have just disposed of. Another is the constructibility of regular polygons, which we shall take up in articles **135–138**. A third is the squaring of the circle, that is, the problem of constructing a square with area equal to that of a given circle. Algebraically this is equivalent to the constructibility of the number $\sqrt{\pi}$, which is clearly impossible once it has been proved that π is transcendental.

The fourth problem is the duplication of the cube. The legend is that the citizens of Delos inquired of the oracle at Delphi what could be done to end the terrible plague decimating their city and received the answer, "Double the size of the altar of Apollo." They replaced the cubical altar with a new one whose sides were twice the length of the sides of the original, but still the plague reigned. A second consultation of the oracle revealed that the requirement was to double the volume of the original altar. Of course this is equivalent to the problem of constructing $\sqrt[3]{2}$, which is not possible with straightedge and compass because the minimal polynomial for $\sqrt[3]{2}$ over the rational field **Q** is $x^3 - 2$, and $[\mathbf{Q}(\sqrt[3]{2}) : \mathbf{Q}] = 3$.

Galois Theory

The Galois theory of equations is one of the most beautiful parts of mathematics and one of the roots of modern algebra. The basic idea of Galois theory is that for a given field, every extension field of a certain kind has associated with it a group, whose structure reveals information about the extension. In particular, splitting fields of polynomials have this property, and solvability of the associated group determines solvability of the polynomial in radicals. Consequently, to prove that equations of the fifth degree are not always solvable in radicals (over the rational field Q), we have only to find a polynomial equation whose splitting field is associated with the symmetric group S_5, which we know is not a solvable group.

This elegant theory is the work of the tormented genius, Évariste Galois (1811–1832), whose brief life is the most tragic episode in the history of mathematics. Persecuted by stupid teachers, twice refused admission to the École Polytechnique, his manuscripts rejected, or even worse, lost by the learned societies, Galois in bitterness immersed himself in the radical politics of the revolution of 1830 and was imprisoned. Upon his release he got involved in a duel and was fatally wounded, dying before his twenty-first birthday. His manuscripts, hastily scribbled in prison and on the eve of his duel, did not receive the attention they deserved until they were read by Liouville in 1846. Only in 1962 was the critical edition of all Galois's writings finally published, but his reputation as a genius of incredible power has been secure for over a hundred years.

Automorphisms

---

122. An *automorphism* of a field E is a one-to-one onto mapping, $\phi: E \to E$, which preserves addition and multiplication, that is,

$$\phi(\alpha + \beta) = \phi\alpha + \phi\beta \quad \text{and} \quad \phi(\alpha\beta) = (\phi\alpha)(\phi\beta).$$

In other words, ϕ is an automorphism of the additive group structure and ϕ^*, the restriction of ϕ to E^*, is an automorphism of the multiplicative group structure.

If ϕ and ψ are automorphisms of the field E, then so is their composition $\phi\psi$. The inverse of an automorphism is again an automorphism. It is easy to see that the set of automorphisms of a field E is a group, which we denote $\mathscr{G}(E)$.

Of course the identity mapping 1_E is an automorphism of the field E, and it is the identity element of the group $\mathscr{G}(E)$.

Whenever we speak of a group of automorphisms of a field E, we shall understand that the group product is composition of automorphisms. In other words, the term "group of automorphisms of E" is synonymous with "subgroup of $\mathscr{G}(E)$."

122α. Show that the groups of automorphisms $\mathscr{G}(\mathbf{Q})$ and $\mathscr{G}(\mathbf{Z}_p)$ are trivial groups.

122β. Determine the group of automorphisms of a field with four elements.

122γ. Determine the group of automorphisms of $\mathbf{Q}(i)$ and $\mathbf{Q}(\sqrt{2})$.

122δ. Prove that the group of automorphisms of $\mathbf{Q}(\zeta)$ where $\zeta = e^{2\pi i/p}$, p prime, is isomorphic to $\mathbf{Z}'_p$.

122ε. Let ϕ be an automorphism of a field E. Prove that the set

$$F = \{\alpha \in E \mid \phi\alpha = \alpha\}$$

is a subfield of E.

122ζ. Let E be a finite field of characteristic p. Show that the mapping $\phi: E \to E$ given by $\phi\alpha = \alpha^p$ is an automorphism. Under what conditions is ϕ an automorphism when E is infinite?

122η. Let $E = F(\alpha)$ and suppose that β is a root in E of a minimal polynomial of α over F. Show that there is a unique automorphism $\phi: E \to E$, such that $\phi\alpha = \beta$ and $\phi c = c$ for $c \in F$.

123. Let ϕ be an automorphism of the field E. We say that ϕ *leaves fixed* an element $\alpha \in E$ if $\phi\alpha = \alpha$. We say that ϕ leaves fixed a subset X of E if $\phi\alpha = \alpha$ for all $\alpha \in X$. The set

$$\{\alpha \in E \mid \phi\alpha = \alpha\}$$

clearly forms a subfield of E, which we call the *fixed field* of ϕ. The fixed field of ϕ is the largest field left fixed by ϕ.

If $\phi_1, \phi_2, \ldots, \phi_n$ are automorphisms of E, then the set

$$\{\alpha \in E \mid \phi_1\alpha = \phi_2\alpha = \cdots = \phi_n\alpha = \alpha\}$$

is called the fixed field of $\phi_1, \phi_2, \ldots, \phi_n$.

As an example, let $\phi: \mathbf{Q}(\sqrt{2}) \to \mathbf{Q}(\sqrt{2})$ be the automorphism given by

$$\phi(a + b\sqrt{2}) = a - b\sqrt{2},$$

where $a, b \in \mathbf{Q}$. Then the fixed field of ϕ is just $\mathbf{Q}$.

For any subset X of a field E the automorphisms of E which leave fixed the set X form a group which we denote $\mathcal{G}(E, X)$.

123α. Let $\zeta = e^{2\pi i/5}$, and let ϕ denote the automorphism of $\mathbf{Q}(\zeta)$ given by $\phi\zeta = \zeta^4$. Prove that the fixed field of ϕ is $\mathbf{Q}(\sqrt{5})$.

123β. Let ϕ be an automorphism of a field E leaving fixed the subfield F. Show that $\alpha \in E$ and a root of $f \in F[x]$ implies $\phi\alpha$ is also a root of f.

123γ. Let ϕ be an automorphism of a field E with fixed field F. Show that ϕ extends uniquely to a mapping

$$\tilde{\phi}: E[x] \to E[x]$$

with the following properties:

   (1) $\tilde{\phi}c = \phi c$ for any constant polynomial c,
   (2) $\tilde{\phi}x = x$,
   (3) $\tilde{\phi}(f + g) = (\tilde{\phi}f) + (\tilde{\phi}g)$,
   (4) $\tilde{\phi}(fg) = (\tilde{\phi}f)(\tilde{\phi}g)$.

Furthermore, show that $\tilde{\phi}f = f$ if and only if $f \in F[x]$.

123δ. Let ϕ be an automorphism of a field E with fixed field F. Suppose that $f \in E[x]$ is monic and splits in E. Prove that if $f\alpha = 0$ always implies $f(\phi\alpha) = 0$ for $\alpha \in E$, then $f \in F[x]$.

123ε. Let E be the splitting field in $\mathbf{C}$ of the polynomial $x^4 + 1$. Find automorphisms of E which have fixed fields $\mathbf{Q}(\sqrt{-2})$, $\mathbf{Q}(\sqrt{2})$, and $\mathbf{Q}(i)$. Is there an automorphism of E whose fixed field is $\mathbf{Q}$?

124.   A finite set of automorphisms of a field E, $\{\phi_1, \phi_2, \ldots, \phi_n\}$, is called *linearly dependent* over E if there are elements of E, $c_1, c_2, \ldots, c_n \in E$, not all zero, such that

$$c_1(\phi_1\alpha) + c_2(\phi_2\alpha) + \cdots + c_n(\phi_n\alpha) = 0$$

for all elements α. Otherwise $\{\phi_1, \phi_2, \ldots, \phi_n\}$ is called *linearly independent*. (Although these definitions are made by analogy with the situation for vector spaces, this analogy is formal and the automorphisms $\phi_1, \phi_2, \ldots, \phi_n$ should not be viewed as elements of a vector space.)

Proposition.   *If $\phi_1, \phi_2, \ldots, \phi_n$ are distinct automorphisms of E, then the set $\{\phi_1, \phi_2, \ldots, \phi_n\}$ is linearly independent over E.*

Proof.   Suppose $\{\phi_1, \phi_2, \ldots, \phi_n\}$ is linearly dependent over E. Among all the relations of linear dependence involving the ϕ_i's, there is a shortest one (that is, one with the fewest nonzero coefficients). Renumbering if necessary, we may assume that such a shortest relation has the form

$$c_1(\phi_1\alpha) + c_2(\phi_2\alpha) + \cdots + c_r(\phi_r\alpha) = 0, \tag{1}$$

for all $\alpha \in E$, where $r \leq n$, and $c_1, c_2, \ldots, c_r$ are nonzero (the zero coefficients have been deleted). Choose an element $\beta \in E$ such that $\phi_1\beta \neq \phi_r\beta$. (The ϕ_i's are distinct by hypothesis.) Now we have

$$c_1(\phi_1\beta)(\phi_1\alpha) + c_2(\phi_2\beta)(\phi_2\alpha) + \cdots + c_r(\phi_r\beta)(\phi_r\alpha) = 0, \tag{2}$$

$$c_1(\phi_r\beta)(\phi_1\alpha) + c_2(\phi_r\beta)(\phi_2\alpha) + \cdots + c_r(\phi_r\beta)(\phi_r\alpha) = 0, \tag{3}$$

for all $\alpha \in E$. Here (2) is obtained by substituting $\beta\alpha$ for α in (1) and observing that $\phi_i(\beta\alpha) = \phi_i(\beta)\phi_i(\alpha)$. Equation (3) is the result of multiplying (1) by $\phi_r\beta$. Subtracting (3) from (2) gives the new and shorter relation

$$c'_1(\phi_1\alpha) + c'_2(\phi_2\alpha) + \cdots + c'_{r-1}(\phi_{r-1}\alpha) = 0 \tag{4}$$

for all α where $c'_i = c_i(\phi_i\beta - \phi_r\beta)$. Then $c'_1 = c_1(\phi_1\beta - \phi_r\beta) \neq 0$, and relation (4) is nontrivial. This is a contradiction, since (1) is the shortest relation, and it establishes the linear independence of $\{\phi_1, \phi_2, \ldots, \phi_n\}$.

125.   *Proposition.*   *If $\phi_1, \phi_2, \ldots, \phi_n$ are distinct automorphisms of E, each of which leaves fixed the subfield F of E, then $[E:F] \geq n$.*

Proof.   Suppose $[E:F] = r < n$ and that $\{\omega_1, \omega_2, \ldots, \omega_r\}$ is a basis for E over F. Consider the system of equations with coefficients in E,

$$\begin{cases} (\phi_1\omega_1)x_1 + (\phi_2\,\omega_1)x_2 + \cdots + (\phi_n\,\omega_1)x_n = 0, \\ (\phi_1\omega_2)x_1 + (\phi_2\,\omega_2)x_2 + \cdots + (\phi_n\,\omega_2)x_n = 0, \\ \quad\vdots \qquad\qquad \vdots \qquad\qquad\qquad \vdots \\ (\phi_1\omega_r)x_1 + (\phi_2\,\omega_r)x_2 + \cdots + (\phi_n\,\omega_r)x_n = 0. \end{cases} \qquad (*)$$

Since this system has fewer equations than unknowns, there is a *nontrivial* solution $x_1 = c_1$, $x_2 = c_2$, $\ldots$, $x_n = c_n$. (That is, not all the c_i's are zero.) Now we show that this implies that $\{\phi_1, \phi_2, \ldots, \phi_n\}$ is linearly dependent— in fact we show that

$$c_1(\phi_1\alpha) + c_2(\phi_2\,\alpha) + \cdots + c_n(\phi_n\,\alpha) = 0$$

for all $\alpha \in E$. Since $\{\omega_1, \omega_2, \ldots, \omega_r\}$ is a basis for E over F, we can write any $\alpha \in E$ as

$$\alpha = a_1\omega_1 + a_2\omega_2 + \cdots + a_r\omega_r$$

for unique $a_1, a_2, \ldots, a_r \in F$. Now we have $\phi_i(a_j) = a_j$, and therefore,

$$\sum_{i=1}^{n} c_i(\phi_i\,\alpha) = \sum_{i=1}^{n} c_i\left(\phi_i\left(\sum_{j=1}^{r} a_j\omega_j\right)\right)$$

$$= \sum_{i=1}^{n} \sum_{j=1}^{r} c_i a_j(\phi_i\,\omega_j)$$

$$= \sum_{j=1}^{r} a_j\left(\sum_{i=1}^{n} c_i(\phi_i\,\omega_j)\right).$$

Since $c_1, c_2, \ldots, c_n$ give a solution of the system (*), it follows that

$$\sum_{i=1}^{n} c_i(\phi_i\,\omega_j) = 0 \quad \text{for} \quad j = 1, 2, \ldots, r.$$

Thus, $\sum_{i=1}^{n} c_i(\phi_i\,\alpha) = 0$, and $\{\phi_1, \phi_2, \ldots, \phi_n\}$ are dependent. Since our hypothesis includes that $\phi_1, \phi_2, \ldots, \phi_n$ are distinct, this contradicts **124**. Consequently, our assumption that $[E:F] < n$ is incorrect, and therefore, $[E:F] \geq n$.

N.B. The hypothesis is only that F is left fixed by each ϕ_i, *not* that F is the fixed field of $\phi_1, \phi_2, \ldots, \phi_n$.

126. Proposition. *If F is the fixed field of a finite group G of automorphisms of E, then $[E:F] = o(G)$.*

Proof. Let $G = \{\phi_1, \phi_2, \ldots, \phi_n\}$ and suppose $[E:F] = r > n$. Let $\{\omega_1, \omega_2, \ldots, \omega_r\}$ be a basis for E over F. Consider the system of equations over E,

$$\begin{cases} (\phi_1\omega_1)x_1 + (\phi_1\omega_2)x_2 + \cdots + (\phi_1\omega_r)x_r = 0, \\ (\phi_2\omega_1)x_1 + (\phi_2\omega_2)x_2 + \cdots + (\phi_2\omega_r)x_r = 0, \\ \vdots \qquad\qquad \vdots \qquad\qquad\qquad \vdots \\ (\phi_n\omega_1)x_1 + (\phi_n\omega_2)x_2 + \cdots + (\phi_n\omega_r)x_r = 0. \end{cases} \tag{$*$}$$

Since this system has $r - n$ more unknowns than equations, there is a nontrivial solution $x_1 = c_1, x_2 = c_2, \ldots, x_r = c_r$ in which $r - n$ of the c_i's may be chosen arbitrarily. For $i = 1, 2, \ldots, r$ let

$$a_i = \phi_1 c_i + \phi_2 c_i + \cdots + \phi_n c_i.$$

We may choose $c_1, c_2, \ldots, c_{r-n}$ so that $a_1, a_2, \ldots, a_{r-n}$ are nonzero. (Why?) For $i = 1, 2, \ldots, r$ the elements a_i are left fixed by each element of G, and consequently, $a_1, a_2, \ldots, a_r \in F$, the fixed field. Now we have

$$\sum_{i=1}^{r} a_i \omega_i = \sum_{i=1}^{r} \left(\sum_{j=1}^{n} \phi_j c_i\right)\omega_i = \sum_{j=1}^{n} \phi_j\left(\sum_{i=1}^{r} c_i(\phi_j^{-1}\omega_i)\right) = 0$$

since $\sum_{i=1}^{r} c_i(\phi_j^{-1}\omega_i) = 0$. This contradicts the linear independence of $\{\omega_1, \omega_2, \ldots, \omega_r\}$. It must be that $[E:F] \leq o(G)$. On the other hand, we know from **125** that $[E:F] \geq o(G) = n$. Thus $[E:F] = o(G)$.

126α. Let E denote the splitting field in $\mathbf{C}$ of $x^4 + 1$ over $\mathbf{Q}$. Prove that $[E:\mathbf{Q}] = 4$.

126β. Find a group of automorphisms of $\mathbf{Q}(\zeta)$, where $\zeta = e^{2\pi i/5}$, of which the fixed field is $\mathbf{Q}$, and determine $[\mathbf{Q}(\zeta):\mathbf{Q}]$. How else can $[\mathbf{Q}(\zeta):\mathbf{Q}]$ be found?

126γ. Let E denote the splitting field in $\mathbf{C}$ of $x^3 - 2$ over $\mathbf{Q}$. Find a group of six automorphisms of E with fixed field $\mathbf{Q}$, thereby showing that $[E:\mathbf{Q}] = 6$.

Galois Extensions

127. A field E is a *Galois extension* of F if F is the fixed field of a finite group of automorphisms of E, which we call the *Galois group* of E over F and denote $\mathscr{G}(E/F)$. With this definition we may restate succinctly the proposition of **126**: *the degree of a Galois extension is the order of its Galois group.* In other words, when E is a Galois extension of F, we have $[E:F] = o(\mathscr{G}(E/F))$. It follows that the Galois group $\mathscr{G}(E/F)$ contains every automorphism of E which leaves F fixed: were there one it did not contain, then by **125** we would have $[E:F] > o(\mathscr{G}(E/F))$, contradicting **126**.

Remark. Some authors use "normal extension" in place of "Galois extension." This is unfortunate, since "normal extension" has another more generally accepted use. (See **129α**.)

127α. Show that an extension of degree 2 is Galois except possibly when the characteristic is 2. Can a field of characteristic 2 have a Galois extension of degree 2?

127β. Show that $\mathbf{Q}(\zeta)$, where $\zeta = e^{2\pi i/5}$, is a Galois extension of $\mathbf{Q}$.

127γ. Show that $\mathbf{Q}(\sqrt[3]{2})$ is not a Galois extension of $\mathbf{Q}$. Find a Galois extension of $\mathbf{Q}$ which contains $\mathbf{Q}(\sqrt[3]{2})$ as a subfield.

127δ. Suppose that $E = F(\alpha)$ is a Galois extension of F. Show that

$$fx = (x - \phi_1\alpha)(x - \phi_2\alpha)\cdots(x - \phi_n\alpha)$$

is a minimal polynomial for α over F, where $\phi_1, \phi_2, \ldots, \phi_n$ are the elements of $\mathscr{G}(E/F)$.

127ε. Let E be a Galois extension of F and suppose that $\alpha \in E$ is an element left fixed only by the identity automorphism of E. Prove that $E = F(\alpha)$.

128. Proposition. *Let $\phi_1, \phi_2, \ldots, \phi_n$ be distinct automorphisms of a field E, each leaving fixed the subfield F. If $[E:F] = n$, then E is a Galois extension of F with group*

$$\mathscr{G}(E/F) = \{\phi_1, \phi_2, \ldots, \phi_n\}.$$

Proof. Since $\{\phi_1, \phi_2, \ldots, \phi_n\} \subset \mathscr{G}(E)$, to show that this set is a group under composition we need only verify that it is a subgroup of $\mathscr{G}(E)$. Suppose that a composition $\phi_i\phi_j \notin \{\phi_1, \phi_2, \ldots, \phi_n\}$. Then $\phi_i\phi_j$ in addition to $\phi_1, \phi_2, \ldots, \phi_n$, leaves F fixed, which by **125** implies $[E:F] \geq n+1$, a contradiction. A similar contradiction arises if $\phi_i^{-1} \notin \{\phi_1, \phi_2, \ldots, \phi_n\}$. Consequently, $\{\phi_1, \phi_2, \ldots, \phi_n\}$ is a subgroup of $\mathscr{G}(E)$, and hence a group. The fixed field F' of $\{\phi_1, \phi_2, \ldots, \phi_n\}$ contains F and satisfies $[E:F'] = n$ by **126**. The equation

$$[E:F] = [E:F'][F':F]$$

yields $[F':F] = 1$, which means $F = F'$. This proves the proposition.

128α.   Let E be an extension of $\mathbf{Z}_p$ such that $[E : \mathbf{Z}_p] = n$. Since E^* is a cyclic group (**100**) and has a generator θ, we know that $E = \mathbf{Z}_p(\theta)$. Let $\phi : E \to E$ be given by $\phi x = x^p$. Prove that $1, \phi, \phi^2, \ldots, \phi^{n-1}$ are distinct automorphisms of E leaving $\mathbf{Z}_p$ fixed and conclude that E is a Galois extension of $\mathbf{Z}_p$ with cyclic Galois group

$$\mathscr{G}(E/\mathbf{Z}_p) = \{1, \phi, \phi^2, \ldots, \phi^{n-1}\}.$$

129.   Theorem.   *E is a Galois extension of F if and only if the following conditions hold:*

(1) *an irreducible polynomial over F of degree m with at least one root in E has m distinct roots in E;*
(2) *E is a simple algebraic extension of F, that is, $E = F(\theta)$ for some element $\theta \in E$ which is algebraic over F.*

Proof.   *Necessity of condition* (1).   Suppose E is a Galois extension of F with group $\mathscr{G}(E/F) = \{\phi_1, \phi_2, \ldots, \phi_n\}$. Let f be a polynomial irreducible over F with a root $\alpha \in E$. We let $\alpha_1, \alpha_2, \ldots, \alpha_r$ denote the distinct values among the elements $\phi_1 \alpha, \phi_2 \alpha, \ldots, \phi_n \alpha \in E$. Then any automorphism in $\mathscr{G}(E/F)$ simply permutes the elements $\alpha_1, \alpha_2, \ldots, \alpha_r$. It follows that each automorphism of $\mathscr{G}(E/F)$ leaves fixed all the coefficients of the polynomial

$$gx = (x - \alpha_1)(x - \alpha_2) \cdots (x - \alpha_r).$$

Thus, all the coefficients of g lie in the fixed field F, that is, g is a polynomial over F. Since α is among the elements $\alpha_1, \alpha_2, \ldots, \alpha_r$, we have $g\alpha = 0$, and consequently, $f \mid g$. However, g splits in E, and therefore f must split in E. Clearly, the roots of f are all distinct.

Necessity of condition (2).   If F is a finite field, then so is E. (Why?) Therefore the multiplicative group E^* is cyclic (**100**) and has a generator θ. It follows that $E = F(\theta)$. This takes care of the case where F is finite.

Suppose F is infinite. Let $\theta \in E$ be an element left fixed by as few automorphisms of the Galois group $\mathscr{G}(E/F)$ as possible. (Why does such an element exist?) We let

$$\mathscr{G}_\theta = \{\phi \in \mathscr{G}(E/F) \mid \phi\theta = \theta\}.$$

Clearly, $\mathscr{G}_\theta$ is a group. *We claim that $\mathscr{G}_\theta$ consists of the identity automorphism alone, or in other words, that $o(\mathscr{G}_\theta) = 1$.* Suppose $o(\mathscr{G}_\theta) > 1$. Then the fixed field B of $\mathscr{G}_\theta$ is a proper subfield of E since $[E : B] = o(\mathscr{G}_\theta)$. Let $\eta \in E - B$. Then there is at least one automorphism in $\mathscr{G}_\theta$ which does not fix η. Take $\eta_1, \eta_2, \ldots, \eta_r$ to be the distinct values among $\phi_1 \eta, \phi_2 \eta, \ldots, \phi_n \eta$, and similarly, take $\theta_1, \theta_2, \ldots, \theta_s$ to be the distinct values among $\phi_1 \theta, \phi_2 \theta, \ldots, \phi_n \theta$. We may assume $\eta = \eta_1$ and $\theta = \theta_1$. Since F is infinite, we may choose $c \in F$ to differ from all the elements

$$\frac{(\eta_i - \eta_1)}{(\theta_1 - \theta_j)}$$

for $i = 2, 3, \ldots, r$ and $j = 2, 3, \ldots, s$. Let $\zeta = \eta + c\theta = \eta_1 + c\theta_1$. The choice of c insures that $\zeta = \eta_i + c\theta_j$ only when $i = j = 1$. Now if ζ is left fixed by $\phi \in \mathscr{G}(E/F)$, we have

$$\zeta = \phi\zeta = \phi(\eta + c\theta) = (\phi\eta) + c(\phi\theta) = \eta_i + c\theta_j = \eta + c\theta,$$

which implies that $\phi\eta = \eta$ and $\phi\theta = \theta$. Thus ζ is fixed only by the automorphisms of $\mathscr{G}(E/F)$ that fix both η and θ. Since there is at least one automorphism in $\mathscr{G}_\theta$ which does not fix η, it is evident that ζ is left fixed by fewer elements of $\mathscr{G}(E/F)$ than θ. This contradicts the choice of θ and establishes that $\mathscr{G}_\theta$ has order 1. It follows that the elements $\phi_1\theta, \phi_2\theta, \ldots, \phi_n\theta$ are distinct: $\phi_i\theta = \phi_j\theta$ implies $\phi_j^{-1}\phi_i\theta = \theta$, from which we infer that $\phi_j^{-1}\phi_i = 1_E$ and $\phi_i = \phi_j$. Furthermore, the elements $\phi_1\theta, \phi_2\theta, \ldots, \phi_n\theta$ are all roots of a minimal polynomial for θ over F. (Why?) Therefore we have $[F(\theta): F] \geq n$. However, $F(\theta) \subset E$ and $[E: F] = n$. Consequently, $E = F(\theta)$.

Sufficiency of the conditions. Let E be an extension of F, which satisfies conditions (1) and (2). Then $E = F(\theta)$ and the minimal polynomial of θ has n distinct roots $\theta_1 = \theta, \theta_2, \ldots, \theta_n$, where $n = [E: F]$ is the degree of the minimal polynomial. Since for all i, $F(\theta_i) \subset E$ and $[F(\theta_i): F] = n$, we have $F(\theta_i) = E$. Now we construct n automorphisms of E, $\phi_1, \phi_2, \ldots, \phi_n$, by setting $\phi_i(\theta) = \theta_i$. Since the set $1, \theta, \theta^2, \ldots, \theta^{n-1}$ forms a basis for E over F, each element of E may be written as $g\theta$ where g is a polynomial over F of degree less than n. Now we set $\phi_i(g\theta) = g\theta_i$. It follows (as the reader should verify) that the mappings $\phi_1, \phi_2, \ldots, \phi_n$ are automorphisms of E. Clearly, they leave F fixed, and therefore E is a Galois extension of F with group $\mathscr{G}(E/F) = \{\phi_1, \phi_2, \ldots, \phi_n\}$.

129α. An extension E of a field F is *normal* if every irreducible polynomial over F with a root in E splits in E. Prove that an extension E of F is Galois if and only if it is finite, separable, and normal. (For the definition and properties of separable extensions see exercises **113α–113μ**. See also **114α**.) Since all extensions in characteristic 0 are separable, we may conclude that *a finite extension of a field of characteristic 0 is Galois if and only if it is normal.* This explains the occasional use of the word " normal" for what we call " Galois" extensions.

129β. Give an example of an extension which is finite and separable but not normal.

129γ. Give an example of an extension which is separable and normal but not finite.

129δ. Let E denote the field of rational functions $\mathbf{Z}_p(x)$. (See **98α**.) Let $F = \mathbf{Z}_p(x^p)$. Show that E is a finite normal extension of F, but not separable.

130. The Fundamental Theorem of Galois Theory. *Let E be a Galois extension of the field F. If B is a field between E and F, then E is a Galois extension of B and $\mathscr{G}(E|B)$ is a subgroup of $\mathscr{G}(E|F)$. Furthermore, B is a Galois extension of F if and only if $\mathscr{G}(E|B)$ is a normal subgroup of $\mathscr{G}(E|F)$, in which case $\mathscr{G}(B|F)$ is isomorphic to the quotient group $\mathscr{G}(E|F)/\mathscr{G}(E|B)$.*

Proof. First we show that E is a Galois extension of B. By **129**, $E = F(\theta)$ for some $\theta \in E$. Clearly $E = B(\theta)$ also. If f is irreducible over B with a root $\alpha \in E$, then $f \,|\, g$ where g is a minimal polynomial for α over F. By **129**, g has all its roots (which number deg g) in E, and they are distinct. Because $f \,|\, g$, the same is true of f. Thus, conditions (1) and (2) of **129** hold for E as an extension of B. Consequently, E is a Galois extension of B. Since $F \subset B \subset E$, it is obvious that $\mathscr{G}(E|B)$ is a *subset* of $\mathscr{G}(E|F)$. Both are subgroups of $\mathscr{G}(E)$, hence $\mathscr{G}(E|B)$ is a subgroup of $\mathscr{G}(E|F)$.

Suppose B is a Galois extension of F. Then $B = F(\xi)$ for some $\xi \in B$. If g is a minimal polynomial for ξ over F and deg $g = m$, then g has m distinct roots in B—all the roots it can have. If $\phi \in \mathscr{G}(E|F)$, then $\phi(g\xi) = g(\phi\xi) = 0$, and $\phi\xi$ is a root of g, hence $\phi\xi \in B$. It follows that ϕ maps B into B, since the element ξ generates B over F. Thus for each automorphism $\phi \in \mathscr{G}(E|F)$, its restriction to B, denoted $\phi \,|\, B$, is an automorphism of B. Furthermore, since ϕ fixes F, $\phi | B$ also fixes F, and therefore $\phi \,|\, B \in \mathscr{G}(B|F)$. All this information can be summed up as follows: *there is a group homomorphism h: $\mathscr{G}(E|F) \to \mathscr{G}(B|F)$ given by $h(\phi) = \phi \,|\, B$.* The kernel of h is the subset of $\mathscr{G}(E|F)$ consisting of all automorphisms whose restriction to B is just 1_B. In other words, Ker $h = \mathscr{G}(E|B)$. The kernel of a homomorphism is always a normal subgroup (**65**). Consequently, $\mathscr{G}(E|B)$ is a normal subgroup of $\mathscr{G}(E|F)$.

Suppose on the other hand that we know $\mathscr{G}(E|B)$ to be a normal subgroup of $\mathscr{G}(E|F)$. Then for $\phi \in \mathscr{G}(E|B)$ and $\psi \in \mathscr{G}(E|F)$ we have $\psi^{-1}\phi\psi \in \mathscr{G}(E|B)$, and for $\beta \in B$ we have $\psi^{-1}\phi\psi\beta = \beta$, or $\phi\psi\beta = \psi\beta$. Fixing ψ and letting ϕ run through $\mathscr{G}(E|B)$ shows that $\psi\beta$ belongs to B, the fixed field of $\mathscr{G}(E|B)$. To summarize, $\beta \in B$ and $\psi \in \mathscr{G}(E|F)$ imply $\psi\beta \in B$. We may once again define a homomorphism h: $\mathscr{G}(E|F) \to \mathscr{G}(B, F)$ with Ker $h = \mathscr{G}(E|B)$. (Here $\mathscr{G}(B, F)$ denotes the group of automorphisms of B which leave F fixed; we do not yet know that F is the fixed field of $\mathscr{G}(B, F)$.) By **67** we know that h induces a monomorphism

$$h': \mathscr{G}(E|F)/\mathscr{G}(E|B) \to \mathscr{G}(B, F).$$

Now it follows that

$$[B\colon F] = [E\colon F]/[E\colon B] = o(\mathscr{G}(E|F))/o(\mathscr{G}(E|B)) \le o(\mathscr{G}(B, F)).$$

On the other hand, F is contained in the fixed field of $\mathcal{G}(B, F)$ and therefore $[B:F] \geq o(\mathcal{G}(B, F))$. Consequently, $[B:F] = o(\mathcal{G}(B, F))$. Then **128** implies that B is a Galois extension of F with group $\mathcal{G}(B/F) = \mathcal{G}(B, F)$. Finally, we note that

$$o(\mathcal{G}(E/F)/\mathcal{G}(E/B)) = o(\mathcal{G}(B/F))$$

implies that h' is an isomorphism.

130α. Prove that if E is a Galois extension of F, then there are only a finite number of fields between E and F.

130β. Let $E = \mathbf{Q}(\zeta)$ where $\zeta = e^{2\pi i/7}$. Show that E is a Galois extension of $\mathbf{Q}$ and determine the Galois group. Find all the fields between $\mathbf{Q}$ and E, the subgroup of $\mathcal{G}(E/\mathbf{Q})$ to which they belong, and determine which are Galois extensions of $\mathbf{Q}$.

130γ. Let E be a Galois extension of F and let B_1 and B_2 be two intermediate fields. (That is, $F \subset B_1 \subset E$ and $F \subset B_2 \subset E$.) We say that B_1 and B_2 are *conjugate* if there is an automorphism $\phi \in \mathcal{G}(E/F)$ such that $\phi B_1 = B_2$. Show that B_1 and B_2 are conjugate if and only if the groups $\mathcal{G}(E/B_1)$ and $\mathcal{G}(E/B_2)$ are conjugate subgroups of $\mathcal{G}(E/F)$.

130δ. Let E be a Galois extension of F with $\mathcal{G}(E/F)$ a cyclic group of order n. Prove that the following conditions hold:

(1) For each divisor d of n there exists precisely one intermediate field B with $[E:B] = d$.
(2) If B_1 and B_2 are two intermediate fields, then $B_1 \subset B_2$ if and only if $[E:B_2]$ divides $[E:B_1]$.

130ε. Prove the converse of **130δ**. In other words, show that if (1) and (2) hold for the intermediate fields of a Galois extension E of F, then $\mathcal{G}(E/F)$ is cyclic.

130ζ. Let E be a finite extension of F and let B_1 and B_2 be intermediate fields such that no proper subfield of E contains both B_1 and B_2. Show that if B_1 is a Galois extension of F, then E is a Galois extension of B_2, and that $\mathcal{G}(E/B_2)$ is isomorphic to a subgroup of $\mathcal{G}(B_1/F)$. Show that $B_1 \cap B_2 = F$ implies that $\mathcal{G}(E/B_2) \approx \mathcal{G}(B_1/F)$.

130η. With the same hypotheses as in **130ζ** prove that if B_1 and B_2 are both Galois extensions of F, then E is a Galois extension of F. Show further that when $B_1 \cap B_2 = F$, $\mathcal{G}(E/F) \approx \mathcal{G}(E/B_1) \times \mathcal{G}(E/B_2)$.

130θ. Let $\mathbf{C}(x)$ denote the field of rational functions over $\mathbf{C}$, the field of complex numbers. Consider the six mappings $\phi_i : \mathbf{C}(x) \to \mathbf{C}(x)$ given by

$$\phi_1 : f(x) \to f(x), \qquad \phi_4 : f(x) \to f\left(\frac{x-1}{x}\right),$$

$$\phi_2 : f(x) \to f(1-x), \qquad \phi_5 : f(x) \to f\left(\frac{1}{1-x}\right),$$

$$\phi_3 : f(x) \to f(1/x), \qquad \phi_6 : f(x) \to f\left(\frac{x}{x-1}\right),$$

for any rational function $f(x) \in \mathbf{C}(x)$. Verify that these mappings form a group of automorphisms of $\mathbf{C}(x)$, and determine the fixed field of this group. How many intermediate fields are there?

130ι. Prove that a finite extension of a finite field is Galois with a cyclic Galois group.

131.   Symmetric Polynomials.

We digress briefly to prove a result on polynomials in several variables needed in the next article. For brevity we avoid studied rigor and appeal to intuition. Readers requiring a more thorough discussion will find one in van der Waerden's *Modern Algebra*, Chapter IV, §26.

Let F^n denote the n-fold cartesian product $F \times F \times \cdots \times F$. (Recall that a point of F^n is an n-tuple $(c_1, c_2, \ldots, c_n)$ of elements of F.)

A *polynomial in n variables* over the field F is an expression of the form

$$f(x_1, x_2, \ldots, x_n) = \sum c(v_1, v_2, \ldots, v_n) x_1^{v_1} x_2^{v_2} \cdots x_n^{v_n},$$

where $\sum$ denotes a finite sum, the coefficients $c(v_1, v_2, \ldots, v_n)$ are elements of F, and the exponents $v_1, v_2, \ldots, v_n$ are nonnegative integers. Each term $x_1^{v_1} x_2^{v_2} \cdots x_n^{v_n}$ is called a *monomial*, and its degree is the sum $v_1 + v_2 + \cdots + v_n$. The *degree* of a polynomial is the highest degree among its monomials with nonzero coefficients.

Given a polynomial f in n variables over F and a permutation of n letters, $\pi \in S_n$, we define a new polynomial f^π in n variables over F by setting

$$f^\pi(x_1, x_2, \ldots, x_n) = f(x_{\pi(1)}, x_{\pi(2)}, \ldots, x_{\pi(n)}).$$

For example, suppose

$$f(x_1, x_2, x_3) = x_1^2 + x_2 x_3 \quad \text{and} \quad \pi = (1, 2, 3).$$

Then

$$f^\pi(x_1, x_2, x_3) = f(x_2, x_3, x_1) = x_2^2 + x_3 x_1.$$

A polynomial f in n variables over F is *symmetric* if $f^\pi = f$ for all $\pi \in S_n$. In other words, a symmetric polynomial is one which remains the same under all permutations of its variables. For example,

$$f(x_1, x_2, \ldots, x_n) = x_1^2 + x_2^2 + \cdots + x_n^2$$

is symmetric. The most important symmetric polynomials, as we shall see, are the *elementary symmetric functions* $\sigma_1, \sigma_2, \ldots, \sigma_n$ defined by the equation

$$(X - x_1)(X - x_2) \cdots (X - x_n) = X^n - \sigma_1 X^{n-1} + \sigma_2 X^{n-2} - \cdots + (-1)^n \sigma_n.$$

It follows that

$$\sigma_1(x_1, x_2, \ldots, x_n) = x_1 + x_2 + \cdots + x_n,$$
$$\sigma_2(x_1, x_2, \ldots, x_n) = \sum_{i<j} x_i x_j,$$
$$\vdots$$
$$\sigma_n(x_1, x_2, \ldots, x_n) = x_1 x_2 \cdots x_n.$$

In general, $\sigma_k(x_1, x_2, \ldots, x_n)$ is the sum of all the monomials $x_{i_1} x_{i_2} \cdots x_{i_k}$, where $i_1 < i_2 < \cdots < i_k$. If g is any polynomial in the elementary symmetric functions, then $g(\sigma_1, \sigma_2, \ldots, \sigma_n) = f(x_1, x_2, \ldots, x_n)$, where f is a symmetric polynomial. For example,

$$\sigma_1^2 - 2\sigma_2 = x_1^2 + x_2^2 + \cdots + x_n^2.$$

Theorem. *A symmetric polynomial in n variables over the field F can be written uniquely as a polynomial in the elementary symmetric functions $\sigma_1, \sigma_2, \ldots, \sigma_n$ over F.*

Proof. The proof is by induction on the number of variables. The case $n = 1$ is trivial. Assume the statement is true for polynomials in $n - 1$ (or fewer) variables. The induction step from $n - 1$ to n will be proved by induction on the degree of the polynomial. The case of zero degree is trivial. Suppose that the statement is true for polynomials in n variables of degree less than m (as well as for all polynomials in fewer variables). Given a polynomial f of degree m which is symmetric in n variables, we let f' denote the polynomial in $n - 1$ variables given by

$$f'(x_1, x_2, \ldots, x_{n-1}) = f(x_1, x_2, \ldots, x_{n-1}, 0).$$

Then f' is symmetric and may be written as a polynomial $g(\sigma_1', \sigma_2', \ldots, \sigma_{n-1}')$ in the elementary symmetric functions in $n - 1$ variables, $\sigma_1', \sigma_2', \ldots, \sigma_{n-1}'$. We note that

$$\sigma_i'(x_1, x_2, \ldots, x_{n-1}) = \sigma_i(x_1, x_2, \ldots, x_{n-1}, 0).$$

132α.   Indicate the modifications of the proof of the theorem above which are necessary to prove the following more general theorem.

Let E be the splitting field (in some extension) of a separable polynomial f over the field F. Then E is a Galois extension of F.

Show that if separability of *f* is dropped from the hypotheses, we may still conclude that *E* is a normal extension of *F*. (See **129α** for definition of *normal*.)

132β.   Let *E* be the splitting field in **C** of a polynomial *f* over **Q** with no repeated roots. Show that $\mathscr{G}(E/\mathbf{Q})$ acts transitively (**86**) on the roots of *f* if and only if *f* is irreducible.

132γ.   Let *E* be the splitting field over **Q** of a polynomial of degree *n*. Prove that $o(\mathscr{G}(E/\mathbf{Q}))$ divides *n*!.

132δ.   Let *E* be the splitting field over **Q** of a polynomial of degree 8 which is reducible over **Q** but has no root in **Q**. Show that $[E:\mathbf{Q}] \le 1{,}440$.

133.   In this article we give an example of a Galois extension for which we can compute the Galois group explicitly.

Let *K* denote the splitting field over **Q** of the polynomial $x^4 - 2$, which is clearly irreducible over **Q** by the Eisenstein criterion (**107**). The roots of $x^4 - 2$ are $\pm\sqrt[4]{2}$ and $\pm i \sqrt[4]{2}$. Clearly $K = \mathbf{Q}(\sqrt[4]{2}, i)$, and consequently each automorphism of $\mathscr{G}(K, \mathbf{Q})$ is determined by its values on $\sqrt[4]{2}$ and *i*. All the possibilities are given by Table 6.

Table 6

Automorphism	Value on $\sqrt[4]{2}$	Value on i
e	$\sqrt[4]{2}$	i
σ	$i\sqrt[4]{2}$	i
σ^2	$-\sqrt[4]{2}$	i
σ^3	$-i\sqrt[4]{2}$	i
τ	$\sqrt[4]{2}$	$-i$
$\sigma\tau$	$i\sqrt[4]{2}$	$-i$
$\sigma^2\tau$	$-\sqrt[4]{2}$	$-i$
$\sigma^3\tau$	$-i\sqrt[4]{2}$	$-i$

Thus, the Galois group $\mathscr{G}(K/\mathbf{Q})$ consists of the eight automorphisms e, σ, σ^2, σ^3, τ, $\sigma\tau$, $\sigma^2\tau$, $\sigma^3\tau$ which satisfy the relations $\sigma^4 = e = \tau^2$ and $\tau\sigma\tau = \sigma^3$. Thus $[K:\mathbf{Q}] = 8$. A basis for *K* over **Q** is

$$\{1, \sqrt[4]{2}, (\sqrt[4]{2})^2, (\sqrt[4]{2})^3, i, i\sqrt[4]{2}, i(\sqrt[4]{2})^2, i(\sqrt[4]{2})^3\}.$$

The element $\sigma\tau$ of $\mathscr{G}(K/\mathbf{Q})$ has order 2, since we have

$$(\sigma\tau)(\sigma\tau) = \sigma(\tau\sigma\tau) = \sigma\sigma^3 = e.$$

Therefore $\{e, \sigma\tau\}$ is a subgroup of $\mathscr{G}(K/\mathbf{Q})$. We shall determine the fixed field of this subgroup. An element $\xi \in K$ may be written

$$\xi = c_1 + c_2\sqrt[4]{2} + c_3(\sqrt[4]{2})^2 + c_4(\sqrt[4]{2})^3$$
$$+ c_5 i + c_6 i \sqrt[4]{2} + c_7 i(\sqrt[4]{2})^2 + c_8 i(\sqrt[4]{2})^3.$$

We can compute $\sigma\tau\xi$ directly as

$$\sigma\tau\xi = c_1 + c_2 i \sqrt[4]{2} - c_3(\sqrt[4]{2})^2 - c_4 i(\sqrt[4]{2})^3$$
$$- c_5 i + c_6 \sqrt[4]{2} + c_7 i(\sqrt[4]{2})^2 - c_8(\sqrt[4]{2})^3.$$

If ξ belongs to the fixed field of $\{e, \sigma\tau\}$, then $\sigma\tau\xi = \xi$ and we must have $c_2 = c_6, c_3 = -c_3, c_4 = -c_8, c_5 = -c_5$, or in other words, $c_3 = 0 = c_5$ and

$$\xi = c_1 + c_2(1 + i)\sqrt[4]{2} + c_7 i(\sqrt[4]{2})^2 + c_8(i - 1)(\sqrt[4]{2})^3$$
$$= c_1 + c_2(1 + i)\sqrt[4]{2} + \tfrac{1}{2}c_7(1 + i)^2(\sqrt[4]{2})^2 + \tfrac{1}{2}c_8(1 + i)^3(\sqrt[4]{2})^3.$$

It follows that the fixed field of $\{e, \sigma\tau\}$ is $\mathbf{Q}((1 + i)\sqrt[4]{2})$. This is not a Galois extension of $\mathbf{Q}$, since $\{e, \sigma\tau\}$ is not a normal subgroup of $\mathscr{G}(K/\mathbf{Q})$.

133α. In the example above, justify the implication that $e, \sigma, \sigma^2, \sigma^3, \tau, \sigma\tau, \sigma^2\tau$, and $\sigma^3\tau$ are automorphisms of K that leave $\mathbf{Q}$ fixed.

133β. Determine the Galois group $\mathscr{G}(E/\mathbf{Q})$, where E is the splitting field over $\mathbf{Q}$ of $x^4 + x^2 - 6$.

133γ. Find the Galois group of the smallest Galois extension of $\mathbf{Q}$ containing $\sqrt{2} + \sqrt[3]{2}$.

133δ. Determine the Galois groups which may occur for splitting fields of cubic equations over $\mathbf{Q}$ and give an equation for each case.

134. *The field of n-th roots of unity.* Let E denote the splitting field over $\mathbf{Q}$ of $x^n - 1$. The roots of $x^n - 1$ are the complex numbers

$$1, \zeta, \zeta^2, \ldots, \zeta^{n-1},$$

where $\zeta = e^{2\pi i/n}$. These roots form a group themselves—the group K_n described in **44**. Since E is a splitting field, it is clearly a Galois extension of $\mathbf{Q}$.

Furthermore, ζ is a primitive element; that is, $E = \mathbf{Q}(\zeta)$. Consequently, each automorphism ϕ of the Galois group $\mathscr{G}(E/\mathbf{Q})$ is completely determined by its value on ζ. Since ϕ can only permute the roots of $x^n - 1$, we must have $\phi\zeta = \zeta^k$ for some k such that $1 \leq k < n$. Not every such k will do: if $(k, n) = d$ and $d > 1$, then

$$\phi(\zeta^{n/d}) = (\zeta^{n/d})^k = (\zeta^{k/d})^n = 1 = \phi(1)$$

and ϕ could not be one to one. Therefore if $\phi \in \mathscr{G}(E/\mathbf{Q})$, then $\phi\zeta = \zeta^k$, where $(k, n) = 1$ and ζ^k is a primitive n-th root of unity (**44**). There are $\phi(n)$ primitive n-th roots of unity (where ϕ denotes the Euler totient function of **25**). It follows that $\mathscr{G}(E/\mathbf{Q})$ can contain at most $\phi(n)$ distinct automorphisms, and as a result $[E : \mathbf{Q}] \leq \phi(n)$. To see that $[E : \mathbf{Q}]$ is exactly $\phi(n)$, we prove the following theorem.

Theorem.   *A minimal polynomial over* $\mathbf{Q}$ *for* $\zeta = e^{2\pi i/n}$ *has every primitive n-th root of unity as a root.*

Proof.   We may factor $x^n - 1$ as a product of polynomials which are irreducible over $\mathbf{Q}$ and which have integral coefficients (**106**). One of these factors, call it f, must have ζ as a root. Since f is irreducible, it must be a minimal polynomial for ζ over $\mathbf{Q}$. What is more, because f and all the other factors have integral coefficients, it follows that f is monic (has leading coefficient 1). We note for future reference that this implies

$$fx = (x - \omega_1)(x - \omega_2) \cdots (x - \omega_r),$$

where $\omega_1, \omega_2, \ldots, \omega_r$ are the roots of f.

Let f_k denote the polynomial over $\mathbf{Z}$ given by $f_k x = f(x^k)$. Since f is monic, we may invoke the division theorem for polynomials over $\mathbf{Z}$ (**99α**) to write f_k uniquely as $f_k = q_k f + r_k$, where q_k and r_k are polynomials over $\mathbf{Z}$ and either $\deg r_k < \deg f$ or $r_k = 0$ (which means that $f | f_k$). Next, we observe that r_k depends only on the congruence class of k modulo n. Indeed, if $k \equiv l \bmod n$, we have $\zeta^k = \zeta^l$, and consequently,

$$f(\zeta^k) - f(\zeta^l) = f_k \zeta - f_l \zeta = 0$$

and ζ is a root of $f_k - f_l$, from which we conclude that $f | (f_k - f_l)$ and $r_k = r_l$. Therefore each r_k equals one of the polynomials $r_1, r_2, \ldots, r_n$.

Let v be a natural number exceeding the content (**106**) of all the polynomials $r_1, r_2, \ldots, r_n$. Then it follows that $r_k = 0$ if there exists a natural number $p > v$ such that $p | r_k$ (p divides each coefficient of r_k). Now we claim: *whenever p is prime and $p > v$, then $p | r_p$ and therefore $r_p = 0$, in other words, $f | f_p$.* To establish this claim, we first remark that $p | (f_p - f^p)$ where f^p is f raised to the p-th power. To put it another way, $f_p = f^p + p\lambda$ where λ is a polynomial over **Z**. (Why?) We may write λ uniquely as $\lambda = \xi f + \rho$ where $\deg \rho < \deg f$ or $\rho = 0$. Now we have

$$f_p = q_p f + r_p = (f^{p-1} + p\xi)f + p\rho,$$

and by uniqueness it follows that $r_p = p\rho$. Since by hypothesis $p > v$, $p \mid r_p$ implies $r_p = 0$ and $f \mid f_p$. We have proven our claim.

Now we are able to show that $(k, n) = 1$ implies ζ^k is a root of f. We would be finished if we knew there were a prime p such that $p > v$ and $p \equiv k$ mod n. For then $f \mid f_p$ would imply that ζ is a root of f_p or equivalently, $f_p \zeta = f(\zeta^p) = f(\zeta^k) = 0$. As a matter of fact, by a theorem of Peter Gustav Lejeune Dirichlet (1805–1859), such a prime will always exist, but we are not able to give the proof of this theorem, which is difficult. Fortunately there is an elementary argument which avoids this point.

Let P denote the product of all the primes less than or equal to v *except those dividing k*, and set $l = k + nP$. Certainly $l \equiv k$ mod n and $\zeta^l = \zeta^k$. Furthermore, primes dividing l must all be larger than v since primes less than or equal to v divide either k or nP, but not both. As a result

$$l = p_1 p_2 \cdots p_s,$$

where p_i is prime and $p_i > v$ for $i = 1, 2, \ldots, s$. By our previous argument we know that $f \mid f_{p_i}$ for each p_i in the factorization of l. Since ζ is a root of f and $f \mid f_{p_i}$, it follows that ζ is a root of f_{p_1} or that $f_{p_1}(\zeta) = f(\zeta^{p_1}) = 0$. Since we now have ζ^{p_1} is a root of f and $f \mid f_{p_2}$, it follows that ζ^{p_1} is a root of f_{p_2}, or that

$$f_{p_2}(\zeta^{p_1}) = f(\zeta^{p_1 p_2}) = 0.$$

In s steps of this kind, we obtain $\zeta^{p_1 p_2 \cdots p_s} = \zeta^l = \zeta^k$ is a root of f and the proof is complete.

Corollary. *If E is the field of n-th roots of unity over* **Q**, *then* $[E: \mathbf{Q}] = \phi(n)$ *and the Galois group* $\mathscr{G}(E/\mathbf{Q})$ *is isomorphic to* $\mathbf{Z}'_n$.

Proof. It is clear from the theorem and the discussion preceding it that $[E: \mathbf{Q}] = \phi(n)$. An isomorphism $\mathscr{G}(E/\mathbf{Q}) \to \mathbf{Z}'_n$ is given by $\phi \to [k]_n$ when ϕ is determined by $\phi\zeta = \zeta^k$. If $\phi\zeta = \zeta^k$ and $\psi\zeta = \zeta^l$, then

$$(\phi\psi)\zeta = \phi(\psi\zeta) = \phi(\zeta^l) = \zeta^{kl},$$

and consequently, $\phi\psi \to [kl]_n = [k]_n[l]_n$, which verifies that the mapping is a homomorphism. It is evident that this homomorphism is also a one-to-one correspondence.

It follows from the theorem that a minimal polynomial for ζ over $\mathbf{Q}$ is given by multiplying together all the factors $x - \zeta^k$ where $(k, n) = 1$ and $1 \le k < n$. This polynomial, denoted Φ_n, is called the *n-th cyclotomic polynomial* (cyclotomic means "circle dividing"). Φ_n is a monic polynomial which is irreducible over $\mathbf{Z}$ and $\deg \Phi_n = \phi(n)$. Gauss was the first to show irreducibility of Φ_p for p a prime. Many proofs of irreducibility for this special case (see **107β**) and for the general case have been found. A detailed survey of those given up to 1900 may be found in Ruthinger, *Die Irreducibilitätsbeweis der Kreisteilungsgleichung*, (Inauguraldissertation, Kaiser Wilhelms Universität, 1907). The proof of the theorem above is an adaption by Artin of an argument of Landau which appears in Vol. 29 (1929) of the *Mathematische Zeitschrift*.

134α.   Show that ζ^k is a primitive (n/d)-th root of unity if and only if $(k, n) = d$. Apply this to prove that

$$x^n - 1 = \prod_{d|n} \Phi_d(x).$$

134β.   Use the formula of **134α** to compute $\Phi_n(x)$ for $1 \le n \le 10$.

134γ.   Let m and n be natural numbers such that every prime p dividing m is a divisor of n. Prove that $\Phi_{mn}(x) = \Phi_n(x^m)$. Use this to compute Φ_{24}, Φ_{36}, and Φ_{100}.

134δ.   Let $E = \mathbf{Q}(\zeta)$ where $\zeta = e^{2\pi i/n}$ and n is odd. Show that E contains all the $2n$-th roots of unity.

134ε.   Prove that $\Phi_{2n}(x) = \Phi_n(-x)$ when n is odd.

134ζ.   Show that for any n-th root of unity,

$$1 + \omega + \omega^2 + \cdots + \omega^{n-1} = \begin{cases} n \text{ when } \omega = 1, \\ 0 \text{ when } \omega \ne 1. \end{cases}$$

134η.   Determine $[\mathbf{Q}(\cos 2\pi r): \mathbf{Q}]$ for $r \in \mathbf{Q}$.

134θ.   Prove that the cyclotomic polynomial satisfies

$$\Phi_n(x) = \prod_{d|n} (x^d - 1)^{\mu(n/d)},$$

where μ denotes the Möbius function (**25β**).

134ι.   Using **134α, γ, ε,** and **θ** and the results of **134β**, compute $\Phi_n(x)$ for $11 \le n \le 36$.

135.   Theorem.   *A regular polygon of n sides is constructible with straightedge and compass if and only if $\phi(n)$ is a power of 2.*

Proof. We observe that the construction of a regular polygon of n sides is equivalent to the division of a circle into n equal arcs. Such a division of the unit circle in the complex plane is equivalent to the construction of the n-th roots of unity. Thus, a regular polygon of n sides is constructible with straight-edge and compass if and only if the splitting field E over $\mathbf{Q}$ of the polynomial $x^n - 1$ is a constructible field. By **120** E is constructible only when the number $[E: \mathbf{Q}] = \phi(n)$ is a power of 2.

On the other hand when $\phi(n) = 2^k$, the Galois group $\mathcal{G}(E/\mathbf{Q}) = \mathbf{Z}_n'$ has order 2^k and by **74** is a solvable group. Explicitly, there is a composition series

$$\{e\} = G_0 \subset G_1 \subset \cdots \subset G_k = \mathcal{G}(E/\mathbf{Q})$$

in which $o(G_i) = 2^i$. We let E_i denote the fixed field of E under G_{k-i}. Then we have a tower of fields,

$$\mathbf{Q} = E_0 \subset E_1 \subset \cdots \subset E_k = E,$$

in which each term is a Galois extension of any of the preceding terms. Furthermore,

$$[E_i: E_{i-1}] = [G_{k+1-i}: G_{k-i}] = 2.$$

A finite induction, using the fact that a quadratic extension of a constructible field is constructible (See **120α**), shows that E is constructible.

136. The preceding theorem leads us to determine the values of n for which the Euler totient function $\phi(n)$ is a power of 2. It follows from **25** that n must have the form $2^r p_1 p_2 \cdots p_k$ where $p_1, p_2, \ldots, p_k$ are the distinct odd primes dividing n and where $\phi(p_i)$ is a power of 2 for $i = 1, 2, \ldots, k$. Since $\phi(p_i) = p_i - 1$, we can reduce the problem to the question of finding all primes of the form $2^m + 1$.

We note that $2^m + 1$ is prime only if m itself is a power of 2. In fact if $m = uv$ where v is an odd number, then we have

$$2^m + 1 = (2^u + 1)(2^{u(v-1)} - 2^{u(v-2)} + \cdots - 2^u + 1).$$

In other words, m cannot be divisible by any odd number and must be a power of 2. Our problem is now reduced to the question of finding all the primes of the form $2^{2^q} + 1$. Fermat (1601–1665) conjectured that all the numbers $2^{2^q} + 1$ are prime, and such numbers are frequently called *Fermat numbers*. For $q < 5$ the number $2^{2^q} + 1$ is prime, but in 1732 Euler (1707–1783) discovered that

$$2^{2^5} + 1 = 641 \times 6,700,417.$$

For no value of q above 4 is $2^{2^q} + 1$ known to be prime, and for many values it is known not to be.

Table 7

q	0	1	2	3	4	5
$2^{2^q} + 1$	3	5	17	257	65,537	4,294,967,297

In summary, we can say that a regular polygon of n sides is known to be constructible whenever $n = 2^r \, s_0^{\varepsilon_0} \, s_1^{\varepsilon_1} \cdots s_4^{\varepsilon_4}$ where $r \geq 0$, $\varepsilon_i = 0$ or 1, and $s_i = 2^{2^i} + 1$, $i = 0, 1, 2, 3, 4$.

The constructions of the equilateral triangle ($n = 3$) and the regular pentagon ($n = 5$) were known to the ancient Greeks. The construction of the regular heptadecagon ($n = 17$) is a discovery of Gauss, who requested that a regular heptadecagon be inscribed on his tomb. The construction of the regular 257-gon was carried out by Richelot in 1832. Professor Hermes of Lingren devoted ten years of his life to the construction of the 65,537-gon. His extensive manuscripts reside in the library at Göttingen. Although many valuable works were destroyed in the flooding of this library, a result of bombings during World War II, Professor Hermes's work was untouched.

136α. List the regular polygons of 100 sides or less which are constructible with straightedge and compass.

136β. Suppose that the regular polygons of m sides and n sides are constructible with straightedge and compass. Prove that a regular polygon of $[m, n]$ sides is constructible.

137. The Regular Pentagon. As a concrete illustration of the preceding articles, we take up construction of the regular pentagon.

Let E denote the splitting field over $\mathbf{Q}$ of the polynomial

$$x^5 - 1 = (x - 1)(x^4 + x^3 + x^2 + x + 1).$$

$E = \mathbf{Q}(\zeta)$ where $\zeta = e^{2\pi i/5} = \cos 72° + i \sin 72°$. The Galois group $\mathscr{G}(E/\mathbf{Q})$ is a group with $[E : \mathbf{Q}] = 4$ elements $\sigma_1, \sigma_2, \sigma_3, \sigma_4$, each of which is completely determined by its value on ζ; $\sigma_i \zeta = \zeta^i$. The group $\mathscr{G}(E/\mathbf{Q})$ is actually a cyclic group of order 4 generated by σ_2. If we write σ for σ_2, then $\sigma^2 = \sigma_4$, $\sigma^3 = \sigma_3$, and $\sigma^4 = \sigma_1$, which is the identity. In other words, $\mathscr{G}(E/\mathbf{Q}) = \{1, \sigma, \sigma^2, \sigma^3\}$ where $\sigma\zeta = \zeta^2$.

The only proper, nontrivial subgroup of $\mathscr{G}(E/\mathbf{Q})$ is the normal subgroup $H = \{1, \sigma^2\}$. Consequently, if B is a field between $\mathbf{Q}$ and E, then by the fundamental theorem of Galois theory, $\mathscr{G}(E/B)$ is a subgroup of $\mathscr{G}(E/\mathbf{Q})$, and in this case it must be that $\mathscr{G}(E/B) = H$ (unless $B = \mathbf{Q}$ or $B = E$). This means that we can determine B as the fixed field of the group H.

The numbers ζ, ζ^2, ζ^3, ζ^4 form a basis for E over $\mathbf{Q}$. (Why?) An element $\alpha = a_1\zeta + a_2\zeta^2 + a_3\zeta^3 + a_4\zeta^4$ of E belongs to B if and only if $\sigma^2\alpha = \alpha$. We compute

$$\sigma^2\alpha = a_1\zeta^4 + a_2\zeta^3 + a_3\zeta^2 + a_4\zeta.$$

Thus, $\sigma^2\alpha = \alpha$ if and only if $a_1 = a_4$ and $a_2 = a_3$. In other words, $\alpha \in B$ if and only if $\alpha = b_1\eta_1 + b_2\eta_2$, where $\eta_1 = \zeta + \zeta^4$ and $\eta_2 = \zeta^2 + \zeta^3$. We note that

$$\eta_1 + \eta_2 = \zeta + \zeta^2 + \zeta^3 + \zeta^4 = -1,$$
$$\eta_1\eta_2 = (\zeta + \zeta^4)(\zeta^2 + \zeta^3) = \zeta^3 + \zeta^4 + \zeta + \zeta^2 = -1.$$

(This follows from the fact that ζ is a root of $x^4 + x^3 + x^2 + x + 1$, and consequently, $\zeta^4 + \zeta^3 + \zeta^2 + \zeta + 1 = 0$.) We now see that η_1 and η_2 are roots of the polynomial over $\mathbf{Q}$,

$$(x - \eta_1)(x - \eta_2) = x^2 - (\eta_1 + \eta_2)x + \eta_1\eta_2 = x^2 + x - 1.$$

Solving this quadratic and noting the position of the roots of unity ζ, ζ^2, ζ^3, ζ^4 on the unit circle (Figure 14), we see that

$$\eta_1 = -\tfrac{1}{2} + \tfrac{1}{2}\sqrt{5} = 2\cos 72°,$$
$$\eta_2 = -\tfrac{1}{2} - \tfrac{1}{2}\sqrt{5} = -2\sin 72°.$$

It follows that $B = \mathbf{Q}(\sqrt{5})$. For the sake of completeness we observe that ζ is a root of the equation $x^2 - \eta_1 x + 1 = 0$ over B.

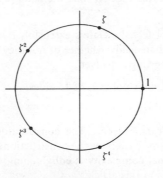

Figure 14

Let us see how we may use this information to construct a regular pentagon. To construct ζ, it is sufficient to construct $\cos 72° = \eta_1/2$. This is easily accomplished as shown in Figure 15.

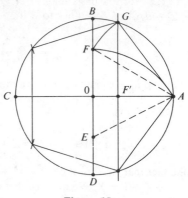

Figure 15

Beginning with the unit circle centered at the origin $O = (0, 0)$ and the four points $A = (1, 0)$, $B = (0, 1)$, $C = (-1, 0)$, and $D = (0, -1)$, the midpoint $E = (0, -1/2)$ of the radius OD is constructed. The line segment AE has length $\sqrt{5}/2$. Next, the point F on OB is constructed so that EF has the same length as AE. Then the length of OF is

$$\eta_1 = -\tfrac{1}{2} + \tfrac{1}{2}\sqrt{5}.$$

The point $G = (\cos 72°, \sin 72°)$ is determined by the perpendicular to OA through F', a point whose distance from O is $\cos 72° = \eta_1/2$. (In a simpler construction one observes that AF has the same length as one side of the pentagon.)

137α.   Analyze completely the Galois extension $E = \mathbf{Q}(\zeta)$ where $\zeta = e^{2\pi i/7}$; determine all intermediate fields, whether or not they are Galois extensions of $\mathbf{Q}$, and all the Galois groups involved.

137β.   Construct a regular polygon of 15 sides.

138.   The Regular Heptadecagon.   The construction of the regular heptadecagon follows the same pattern as the construction of the regular pentagon. However, the additional complexity is enlightening, and we shall sketch the algebraic preliminaries in this article.

Let E denote the splitting field over $\mathbf{Q}$ of the polynomial

$$x^{17} - 1 = (x - 1)(x^{16} + x^{15} + \cdots + x + 1).$$

The roots of $x^{17} - 1$ are the complex numbers $1, \zeta, \zeta^2, \ldots, \zeta^{16}$, where $\zeta = e^{2\pi i/17}$. $E = \mathbf{Q}(\zeta)$ and $[E: \mathbf{Q}] = 16$.

The Galois group $\mathscr{G}(E/\mathbf{Q})$, which is isomorphic to $\mathbf{Z}'_{17}$, may be represented as a set with 16 elements, $\phi_1, \phi_2, \ldots, \phi_{16}$, where ϕ_i is the automorphism of E determined by $\phi_i \zeta = \zeta^i$. This group is a cyclic group of order 16 generated by ϕ_3 as we shall see. From now on we shall denote ϕ_3 simply by ϕ. Since $\phi\zeta = \zeta^3$, we have

$$(\phi^2)\zeta = \phi(\phi\zeta) = \phi(\zeta^3) = (\phi\zeta)^3 = (\zeta^3)^3 = \zeta^{3^2},$$

and, in general, $(\phi^i)\zeta = \zeta^{3^i}$. However, $\zeta^{17} = 1$ and the value of ζ^k depends only on the congruence class of k modulo 17. At this point we need a table of powers of 3 modulo 17.

Table 8

i	0	1	2	3	4	5	6	7	8	9	10	11	12	13	14	15
$3^i \pmod{17}$	1	3	9	10	13	5	15	11	16	14	8	7	4	12	2	6

From this table we can see that $\phi^0 = \phi_1$, $\phi^1 = \phi_3$, $\phi^2 = \phi_9$, and so forth; and we can verify that the powers of ϕ, that is, $1 = \phi^0, \phi^1, \phi^2, \ldots, \phi^{15}$, run through the set $\phi_1, \phi_2, \ldots, \phi_{16}$. It also serves to interpret ϕ_j as a power of ϕ.

In what follows we let ζ_i denote $\phi^i\zeta = \zeta^{3^i}$. We note that

$$\phi^j\zeta_i = \phi^j(\phi^i\zeta) = \phi^{j+i}\zeta = \zeta_{i+j}.$$

This fact is very convenient for making computations. (Table 8 may be used to convert ζ_k to a power of ζ and vice versa.)

The group $\mathscr{G}(E/\mathbf{Q}) = \{1, \phi, \phi^2, \ldots, \phi^{15}\}$ has three proper, nontrivial subgroups, each of which is normal, since $\mathscr{G}(E/\mathbf{Q})$ is abelian. These subgroups are

$$
\begin{aligned}
G_1 &= \{1, \phi^8\} && \text{order 2,} \\
G_2 &= \{1, \phi^4, \phi^8, \phi^{12}\} && \text{order 4,} \\
G_3 &= \{1, \phi^2, \phi^4, \ldots, \phi^{14}\} && \text{order 8.}
\end{aligned}
$$

In fact, the series

$$\{1\} = G_0 \subset G_1 \subset G_2 \subset G_3 \subset G_4 = \mathscr{G}(E/\mathbf{Q})$$

is the one and only composition series which $\mathscr{G}(E/\mathbf{Q})$ admits.

By the fundamental theorem of Galois theory there are three intermediate fields: the fixed fields B_1, B_2, and B_3 of G_1, G_2, and G_3, respectively. We have a tower of fields

$$\mathbf{Q} \subset B_3 \subset B_2 \subset B_1 \subset E.$$

We note that the complex numbers $\zeta, \zeta^2, \ldots, \zeta^{16}$ form a basis for E over $\mathbf{Q}$, or in other words, the numbers $\zeta_0, \zeta_1, \ldots, \zeta_{15}$ form a basis. Furthermore,

$$\zeta + \zeta^2 + \cdots + \zeta^{16} = \zeta_0 + \zeta_1 + \cdots + \zeta_{15} = -1.$$

(Why?) With these facts at our disposal we are ready to determine the fields B_1, B_2, and B_3.

B_3 is the fixed field of $G_3 = \{1, \phi^2, \phi^4, \ldots, \phi^{14}\}$. Since G_3 is a cyclic group generated by ϕ^2, it follows that $\alpha \in E$ is fixed by every element of G_3 if and only if it is fixed by ϕ^2. In other words,

$$B_3 = \{\alpha \in E \mid \phi^2 \alpha = \alpha\}.$$

Using the basis $\zeta_0, \zeta_1, \ldots, \zeta_{15}$, we write

$$\alpha = a_0 \zeta_0 + a_1 \zeta_1 + \cdots + a_{15} \zeta_{15}, \qquad a_i \in \mathbf{Q}.$$

Then

$$\phi^2 \alpha = a_0 \zeta_2 + a_1 \zeta_3 + \cdots + a_{13} \zeta_{15} + a_{14} \zeta_0 + a_{15} \zeta_1.$$

(Note that $\phi^2 \zeta_{14} = \zeta_{16} = \zeta^{3^{16}} = \zeta = \zeta_0$ and $\phi^2 \zeta_{15} = \zeta_1$.) Now we see that $\phi^2 \alpha = \alpha$ and $\alpha \in B_3$ if and only if

$$a_0 = a_2 = a_4 = \cdots = a_{14},$$

and

$$a_1 = a_3 = a_5 = \cdots = a_{15}.$$

Thus, $\alpha \in B_3$ if and only if $\alpha = a_0 \eta_0 + a_1 \eta_1$ where $a_0, a_1 \in \mathbf{Q}$ and

$\eta_0 = \zeta_0 + \zeta_2 + \zeta_4 + \cdots + \zeta_{14} = \zeta + \zeta^9 + \zeta^{13} + \zeta^{15} + \zeta^{16} + \zeta^8 + \zeta^4 + \zeta^2,$
$\eta_1 = \zeta_1 + \zeta_3 + \zeta_5 + \cdots + \zeta_{15} = \zeta^3 + \zeta^{10} + \zeta^5 + \zeta^{11} + \zeta^{14} + \zeta^7 + \zeta^{12} + \zeta^6.$

Now

$$\eta_0 + \eta_1 = \zeta_0 + \zeta_1 + \cdots + \zeta_{15} = -1.$$

Furthermore, $\eta_0 \eta_1$ is left fixed by ϕ^2 and must be expressible as $a_0 \eta_0 + a_1 \eta_1$ for some $a_0, a_1 \in \mathbf{Q}$. Multiplying the expressions for η_0 and η_1 yields

$$\eta_0 \eta_1 = 4\eta_0 + 4\eta_1 = -4.$$

Thus, η_0 and η_1 are roots of the polynomial $x^2 + x - 4$, and $B_3 = \mathbf{Q}(\eta)$ where η is either root η_0 or η_1 of the polynomial. Of course we can solve the equation $x^2 + x - 4 = 0$ to determine

$$\eta = -\tfrac{1}{2} \pm \tfrac{1}{2}\sqrt{17},$$

from which it follows that $B_3 = \mathbf{Q}(\sqrt{17})$. The elements η_0 and η_1 are called *periods* of length 8, since each is a sum of eight of the roots $\zeta_0, \zeta_1, \ldots, \zeta_{15}$.

The fields B_2 and B_1 are determined in a similar fashion. A basis for B_2 over $\mathbf{Q}$ consists of the four periods of length 4,

$$\begin{aligned}
\xi_0 &= \zeta_0 + \zeta_4 + \zeta_8 + \zeta_{12}, \\
\xi_1 &= \zeta_1 + \zeta_5 + \zeta_9 + \zeta_{13}, \\
\xi_2 &= \zeta_2 + \zeta_6 + \zeta_{10} + \zeta_{14}, \\
\xi_3 &= \zeta_3 + \zeta_7 + \zeta_{11} + \zeta_{15}.
\end{aligned}$$

Direct computation shows that the following relations hold among the ξ_i:

$$\begin{aligned}
\xi_0 + \xi_2 &= \eta_0, \qquad \xi_1 + \xi_3 = \eta_1, \\
\xi_0 \xi_2 &= -1 = \xi_1 \xi_3, \\
\xi_1 &= (\xi_0 - 1)/(\xi_0 + 1), \\
\xi_2 &= (\xi_1 - 1)/(\xi_1 + 1), \\
\xi_3 &= (\xi_2 - 1)/(\xi_2 + 1), \\
\xi_0 &= (\xi_3 - 1)/(\xi_3 + 1).
\end{aligned}$$

Thus, ξ_0 and ξ_2 are roots of the polynomial $x^2 - \eta_0 x - 1$ over B_3, and ξ_1 and ξ_3 are roots of $x^2 - \eta_1 x - 1$. Consequently, $B_2 = B_3(\xi)$ where ξ is a root of $x^2 - \eta x - 1$.

A basis for B_1 over $\mathbf{Q}$ consists of eight periods of length 2:

$$\begin{array}{lll}
\lambda_0 = \zeta_0 + \zeta_8, & \lambda_4 = \zeta_4 + \zeta_{12}, & \lambda_0 + \lambda_4 = \xi_0, \\
\lambda_1 = \zeta_1 + \zeta_9, & \lambda_5 = \zeta_5 + \zeta_{13}, & \lambda_1 + \lambda_5 = \xi_1, \\
\lambda_2 = \zeta_2 + \zeta_{10}, & \lambda_6 = \zeta_6 + \zeta_{14}, & \lambda_2 + \lambda_6 = \xi_2, \\
\lambda_3 = \zeta_3 + \zeta_{11}, & \lambda_7 = \zeta_7 + \zeta_{15}, & \lambda_3 + \lambda_7 = \xi_3.
\end{array}$$

At this stage it is easy to show how to multiply these periods:

$$\begin{aligned}
\lambda_0 \lambda_4 &= (\zeta_0 + \zeta_8)(\zeta_4 + \zeta_{12}) = (\zeta + \zeta^{16})(\zeta^{13} + \zeta^4) \\
&= \zeta^{14} + \zeta^5 + \zeta^{29} + \zeta^{20} \\
&= \zeta^{14} + \zeta^5 + \zeta^{12} + \zeta^3 \\
&= \zeta_9 + \zeta_5 + \zeta_{13} + \zeta_1 \\
&= \xi_1.
\end{aligned}$$

Similar computations show $\lambda_1 \lambda_5 = \xi_2$, $\lambda_2 \lambda_6 = \xi_3$, and $\lambda_3 \lambda_7 = \xi_0$. Consequently,

$$\begin{aligned}
&\lambda_0 \text{ and } \lambda_4 \text{ are roots of } x^2 - \xi_0 x + \xi_1 \text{ over } B_2, \\
&\lambda_1 \text{ and } \lambda_5 \text{ are roots of } x^2 - \xi_1 x + \xi_2 \text{ over } B_2, \\
&\lambda_2 \text{ and } \lambda_6 \text{ are roots of } x^2 - \xi_2 x + \xi_3 \text{ over } B_2, \\
&\lambda_3 \text{ and } \lambda_7 \text{ are roots of } x^2 - \xi_3 x + \xi_0 \text{ over } B_2.
\end{aligned}$$

Using the relation between ξ_1 and ξ_0 above, we see that λ_0 and λ_4 are roots of

$$(\xi_0 + 1)x^2 - (\xi_0 + 1)\xi_0 x - (\xi_0 - 1).$$

We conclude (without giving details) that $B_1 = B_2(\lambda)$ where λ is a root of the polynomial

$$(\xi + 1)x^2 - (\xi + 1)\xi x - (\xi - 1)$$

over B_2.

Finally the same kind of analysis shows that $E = B_2(\zeta)$ where ζ is a root of $x^2 - \lambda x + 1$ over B_1.

In summary we have worked out the following relations between the fields $\mathbf{Q}, B_3, B_2, B_1, E$:

$E = B_1(\zeta)$ where ζ is a root of $x^2 - \lambda x + 1$ over B_1,

$B_1 = B_2(\lambda)$ where λ is a root of $(\xi + 1)x^2 - (\xi + 1)\xi x - (\xi - 1)$ over B_2,

$B_2 = B_3(\xi)$ where ξ is a root of $x^2 - \eta x + 1$ over B_3,

$B_3 = \mathbf{Q}(\eta)$ where η is a root of $x^2 + x - 4$ over $\mathbf{Q}$.

We could use this information to formulate a geometric construction of the regular heptadecagon, but there is little interest in actually doing so. Many constructions are available to the reader. (See Eves, *A Survey of Geometry*, Vol. I, p. 217, or Hardy and Wright, *An Introduction to the Theory of Numbers*, p. 57.)

Solvability of Equations by Radicals

---

139.   Let f be a polynomial over a number field F. The equation $fx = 0$ is *solvable by radicals* if all the roots of f can be obtained from elements of F by a finite sequence of rational operations (addition, subtraction, multiplication, and division) and extractions of n-th roots.

For example, the sixth-degree equation over $\mathbf{Q}$,

$$x^6 - 6x^4 + 12x^2 - 15 = (x^2 - 2)^3 - 7 = 0,$$

is solvable by radicals. In fact all six roots may be expressed as $\sqrt[2]{2 + \sqrt[3]{7}}$ provided we interpret $\sqrt[3]{7}$ as *any* of the three cube roots of 7 and $\sqrt[2]{2 + \sqrt[3]{7}}$ as *either* of the square roots of $2 + \sqrt[3]{7}$. Quadratic, cubic, and quartic equations are solvable in radicals.

In 1799 Paolo Ruffini (1765–1822) tried to prove the existence of quintic equations not solvable in radicals. The argument Ruffini gave was inadequate, and the question was settled decisively by Abel in 1824. Galois gave a necessary and sufficient condition for solvability of an equation of any degree by radicals, which dramatically supersedes the work of Abel and Ruffini.

140. The simplest and most clear-cut case of an equation over a number field F which is solvable by radicals is the equation $x^n - \alpha = 0$ where $\alpha \in F^*$. We have already examined the special case $F = \mathbf{Q}$ and $\alpha = 1$ in **134.** Now we take up the general case as preparation for the Galois criterion for solvability by radicals.

Let E denote the splitting field of $x^n - \alpha$ over F. If β is a root of $x^n - \alpha$, then the other roots are $\beta\zeta, \beta\zeta^2, \ldots, \beta\zeta^{n-1}$, where $\zeta = e^{2\pi i/n}$. Since β and $\beta\zeta$ belong to E, it follows that $\zeta = (\beta\zeta)/\beta$ belongs to E, and therefore E contains all the n-th roots of unity, $1, \zeta, \zeta^2, \ldots, \zeta^{n-1}$. Clearly, $E = F(\zeta, \beta)$.

E is a Galois extension of F by **132,** and since $E = F(\zeta, \beta)$, each element of the Galois group $\mathscr{G}(E/F)$ is determined by its value on the two elements ζ and β. If $\phi \in \mathscr{G}(E/F)$, then ϕ must carry ζ to ζ^k where $(k, n) = 1$. (This is shown by the argument of **134.**) On the other hand, ϕ can only permute the roots $\beta, \beta\zeta, \ldots, \beta\zeta^{n-1}$ of $x^n - \alpha$, so that $\phi(\beta) = \beta\zeta^l$. Thus, the two numbers k and l determine the automorphism ϕ completely. In general only certain values of k and l will give elements of $\mathscr{G}(E/F)$.

140α. Determine the Galois group $\mathscr{G}(E/F)$ where E is the splitting field over F of $x^6 - 8$ for the cases when $F = \mathbf{Q}$, $F = \mathbf{Q}(\sqrt{2})$, and $F = \mathbf{Q}(\omega)$ where

$$\omega = e^{2\pi i/3} = -\tfrac{1}{2} + \tfrac{1}{2}\sqrt{-3}.$$

141. **Theorem.** *If E is the splitting field of the polynomial $x^n - \alpha$ over a number field F, then the Galois group $\mathscr{G}(E/F)$ is solvable.*

Proof. By the analysis of **140,** $E = F(\zeta, \beta)$ where $\zeta = e^{2\pi i/n}$ and β is a root of $x^n - \alpha$. Let $B = F(\zeta)$. Then B is the splitting field of $x^n - 1$ over F and is a Galois extension. By the fundamental theorem of Galois theory (**130**) we know that $\mathscr{G}(E/B)$ is a normal subgroup of $\mathscr{G}(E/F)$ and that $\mathscr{G}(B/F)$ is the quotient group. $\mathscr{G}(E/B)$ contains just those automorphisms of $\mathscr{G}(E/F)$ which leave ζ fixed. In terms of **140,** $\phi \in \mathscr{G}(E/B)$ when $k = 1$. Consequently, an automorphism $\phi \in \mathscr{G}(E/B)$ is completely determined by the number l where $\phi(\beta) = \beta\zeta^l$. In fact the assignment $\phi \mapsto l$ identifies $\mathscr{G}(E/B)$ with a subgroup of the finite abelian group $\mathbf{Z}_n$. (Why?) It follows that $\mathscr{G}(E/B)$ is a solvable group

(75). Next we want to see that $\mathcal{G}(B/F)$ may be identified with a subgroup of $\mathbf{Z}'_n$, and this will show that $\mathcal{G}(B/F)$ is solvable. From the proof of the fundamental theorem **(130)**, we recall that the epimorphism $\mathcal{G}(E/F) \to \mathcal{G}(B/F)$ is given by the assignment $\phi \mapsto \phi \,|\, B$. However, the restriction $\phi \,|\, B$ of any $\phi \in \mathcal{G}(E/F)$ is completely determined by the number k of **140**. The mapping given by $(\phi \,|\, B) \mapsto k$ identifies $\mathcal{G}(B/F)$ with a subgroup of $\mathbf{Z}'_n$. Now we have that $\mathcal{G}(E/F)$ is solvable, since the normal subgroup $\mathcal{G}(E/B)$ and the corresponding quotient group $\mathcal{G}(B/F)$ are solvable **(75)**.

142. A *radical tower* over F is a tower of number fields

$$F = F_0 \subset F_1 \subset \cdots \subset F_n$$

in which, for $i = 1, 2, \ldots, n$, F_i is the splitting field of a polynomial $x^{k_i} - \alpha_i$ over F_{i-1}. Such a tower is Galois if the top field F_n is a Galois extension of the ground field F_0.

Proposition.   *Every radical tower can be embedded in a Galois radical tower.*

Proof.   We shall show that given a radical tower over F,

$$F = F_0 \subset F_1 \subset \cdots \subset F_n,$$

we can construct a Galois radical tower over F,

$$F = \tilde{F}_0 \subset \tilde{F}_1 \subset \cdots \subset \tilde{F}_m,$$

such that $F_n \subset \tilde{F}_m$. We begin by setting $\tilde{F}_1 = F_1$. Since F_1 is the splitting field of $x^{k_1} - \alpha_1$ over F_0, by **132** F_1 is a Galois extension of F_0. (If $n = 1$, we would be finished—the two-story tower $\tilde{F}_0 \subset \tilde{F}_1$ is Galois.) Let the Galois group of $\tilde{F}_1$ over $\tilde{F}_0$ be

$$\mathcal{G}(\tilde{F}_1/\tilde{F}_0) = \{\phi_1, \phi_2, \ldots, \phi_r\}.$$

F_2 is the splitting field of $x^{k_2} - \alpha_2$ where $\alpha_2 \in \tilde{F}_1 = F_1$ and α_2 is algebraic over F_1. Now we let $\tilde{F}_2$ be the splitting field of $x^{k_2} - \phi_1\alpha_2$ over $\tilde{F}_1$, $\tilde{F}_3$ the splitting field of $x^{k_2} - \phi_2\alpha_2$ over $\tilde{F}_2$, and so forth, down to $\tilde{F}_{r+1}$, which is the splitting field of $x^{k_2} - \phi_r\alpha_2$ over $\tilde{F}_r$. Now the field $\tilde{F}_{r+1}$ is a Galois extension of F because it is the splitting field of the polynomial

$$(x^{k_1} - \alpha_1)(x^{k_2} - \phi_1\alpha_2)(x^{k_2} - \phi_2\alpha_2) \cdots (x^{k_2} - \phi_r\alpha_2),$$

all of whose coefficients lie in F. (Why?) Furthermore, α_2 is among the numbers $\phi_1\alpha_2, \phi_2\alpha_2, \ldots, \phi_r\alpha_2$, and consequently, $F_2 \subset \tilde{F}_{r+1}$. (If $n = 2$, we are finished.)

This argument can be continued, extending the Galois radical tower

$$F = \tilde{F}_0 \subset \tilde{F}_1 \subset \cdots \subset \tilde{F}_{r+1}$$

to a new one containing F_3. In performing this next stage of the construction, it is necessary to use the Galois group $\mathscr{G}(\tilde{F}_{r+1}/\tilde{F}_0)$ in place of $\mathscr{G}(\tilde{F}_1/\tilde{F}_0)$ and α_3 in place of α_2, but otherwise the argument is similar. The entire argument is iterated until a Galois extension $\tilde{F}_m$ containing F_n is reached.

143. A number field E is a *radical extension* of a number field F if there exists a radical tower over F,

$$F = F_0 \subset F_1 \subset \cdots \subset F_n,$$

such that $E \subset F_n$. In view of the preceding proposition, we may assume without loss of generality that F_n is a Galois extension of F.

Proposition. *If D is a radical extension of E and E is a radical extension of F, then D is a radical extension of F.*

Proof. Let

$$F = F_0 \subset F_1 \subset \cdots \subset F_n$$

and

$$E = E_0 \subset E_1 \subset \cdots \subset E_m$$

be radical towers such that $E \subset F_n$ and $D \subset E_m$. Suppose that E_i is the splitting field of $x^{k_i} - \alpha_i$ over E_{i-1}. Then we let F_{n+i} be the splitting field of $x^{k_i} - \alpha_i$ over F_{n+i-1} in order to define inductively fields $F_{n+1}, F_{n+2}, \ldots, F_{n+m}$. It follows that

$$F = F_0 \subset F_1 \subset \cdots \subset F_n \subset F_{n+1} \subset \cdots \subset F_{n+m}$$

is a radical tower over F. Furthermore, for $i = 0, 1, \ldots, m$, we have $E_i \subset F_{n+i}$. Hence, $D \subset E_m \subset F_{n+m}$ and D is therefore a radical extension of F.

144. Proposition. *If a number field E is a Galois extension of a number field F and the Galois group $\mathscr{G}(E/F)$ is cyclic, then E is a radical extension of F.*

Proof. First we shall prove the proposition for the special case that F contains the n-th roots of unity, $1, \zeta, \ldots, \zeta^{n-1}$ for $n = o(\mathscr{G}(E/F))$. The general case will follow from the special case.

By **129** we have $E = F(\theta)$ for some element θ which is algebraic over F. We define the *Lagrange resolvent* $(\zeta^k, \theta) \in E$ *of* θ *by* ζ^k with the formula

$$(\zeta^k, \theta) = \sum_{i=0}^{n-1} \zeta^{ki}(\phi^i\theta) = \theta + \zeta^k(\phi\theta) + \cdots + \zeta^{k(n-1)}(\phi^{n-1}\theta), \tag{1}$$

where $\mathscr{G}(E/F) = \{1, \phi, \phi^2, \ldots, \phi^{n-1}\}$. Next we compute the sum of all the Lagrange resolvents of θ:

$$\sum_{k=0}^{n-1} (\zeta^k, \theta) = \sum_{k=0}^{n-1} \sum_{i=0}^{n-1} \zeta^{ki}(\phi^i\theta) = \sum_{i=0}^{n-1} \left(\sum_{k=0}^{n-1} (\zeta^i)^k\right)(\phi^i\theta) = n\theta. \tag{2}$$

The last step in this computation is justified by the observation:

$$\sum_{k=0}^{n-1} (\zeta^i)^k = 1 + \zeta^i + \cdots + (\zeta^i)^{n-1} = \begin{cases} n, & \text{for } i \equiv 0 \bmod n, \\ \dfrac{1 - (\zeta^i)^n}{1 - \zeta^i} = 0, & \text{for } i \not\equiv 0 \bmod n. \end{cases} \tag{3}$$

Consequently, we have

$$\theta = \frac{1}{n} \sum_{k=0}^{n-1} (\zeta^k, \theta).$$

Now we observe how the Lagrange resolvents behave under the automorphisms of the Galois group $\mathscr{G}(E/F)$. Since ϕ leaves F fixed, we have $\phi(\zeta^i) = \zeta^i$ for each n-th root of unity. Consequently, we can compute directly that

$$\phi(\zeta^k, \theta) = (\zeta^k, \phi\theta) = \zeta^{-k}(\zeta^k, \theta),$$

and what is more,

$$\phi(\zeta^k, \theta)^n = (\zeta^{-k})^n(\zeta^k, \theta)^n = (\zeta^k, \theta)^n.$$

In other words, ϕ leaves the number $\alpha_k = (\zeta^k, \theta)^n$ fixed, As a result, all the elements of $\mathscr{G}(E/F)$ leave fixed each of the numbers α_k for $k = 0, 1, \ldots, n-1$, and consequently these numbers belong to F. Now we can construct inductively a radical tower

$$F = F_0 \subset F_1 \subset \cdots \subset F_n$$

by taking F_{i+1} to be the splitting field of $x^n - \alpha_i$ over F_i. Then F_n contains all the Lagrange resolvents of θ, and hence,

$$\theta = \frac{1}{n} \sum_{k=0}^{n-1} (\zeta^k, \theta)$$

belongs to F_n. Thus, $E = F(\theta) \subset F_n$, and we have shown that E is a radical extension of F.

The general case, in which F need not contain all the n-th roots of unity, is proved from the special case as follows. Since $E = F(\theta)$ is a Galois extension of F, E is the splitting field of a minimal polynomial for θ over F, call it f. The field $F(\zeta, \theta)$ is the splitting field of the polynomial g over F given by $gx = (x^n - 1)(fx)$. By **132**, $F(\zeta, \theta)$ is a Galois extension of F. It follows from **130** that $F(\zeta, \theta)$ is a Galois extension of $F(\zeta)$. A homomorphism of Galois groups

$$H: \mathscr{G}(F(\zeta, \theta)/F(\zeta)) \to \mathscr{G}(E/F)$$

is defined by $H(\psi) = \psi \,|\, E$ (the restriction of ψ to E). It is not difficult to verify that ψ is well defined, preserves composition of automorphisms, and is one to one. As a result the group $\mathscr{G}(F(\zeta, \theta)/F(\zeta))$ is isomorphic to Im H, which is a subgroup of the cyclic group $\mathscr{G}(E/F)$, and hence, by **43** is itself a cyclic group. Therefore $\mathscr{G}(F(\zeta, \theta)/F(\zeta))$ is a cyclic group, and its order m (which is also the order of Im H) divides n, the order of $\mathscr{G}(E/F)$. Thus $F(\zeta, \theta)$, as an extension of $F(\zeta)$, satisfies the hypothesis of the proposition: its Galois group is cyclic of order m. Furthermore, $F(\zeta)$ contains the m-th roots of unity, $1, \zeta^{n/m}, \zeta^{2n/m}, \ldots, \zeta^{(m-1)n/m}$, because $m \,|\, n$. We have the situation of the special case, and may conclude that $F(\zeta, \theta)$ is a radical extension of $F(\zeta)$. Now $F(\zeta)$ is quite clearly a radical extension of F, and by **143**, so is $F(\zeta, \theta)$. If $F = F_0 \subset F_1 \subset \cdots \subset F_n$ is a radical tower such that F_n contains $F(\zeta, \theta)$, then F_n contains $E = F(\theta)$, which is a subfield of $F(\zeta, \theta)$. Thus E is a radical extension of F.

144α. Let E be a Galois extension of a number field F, with the property that $B_1 \subset B_2$ or $B_2 \subset B_1$ for any two intermediate fields B_1 and B_2. Show that E is a radical extension of F.

144β. Let E be a Galois extension of a number field F with $\mathscr{G}(E/F)$ abelian. Show that E is a radical extension of F.

145. *Theorem* (Galois). *Let f be a polynomial over a number field F and let E be its splitting field. The equation $fx = 0$ is solvable by radicals if and only if the Galois group $\mathscr{G}(E/F)$ is solvable.*

Proof. A moment's reflection reveals the equivalence of the two statements:

(1) the equation $fx = 0$ is solvable by radicals over F;
(2) the splitting field of f is a radical extension of F.

(In fact the second statement is often taken as a definition of the first.) We must prove that E is a radical extension of F if and only if $\mathscr{G}(E/F)$ is solvable. Suppose E is a radical extension of F. Then there is a Galois radical tower

$$F = F_0 \subset F_1 \subset \cdots \subset F_n$$

such that $E \subset F_n$. F_n is a Galois extension of $F = F_0$, and hence, a Galois extension of each F_i. Furthermore, each F_i is a Galois extension of F_{i-1}. Consequently, setting $G_i = \mathscr{G}(F_n/F_{n-i})$, we have that

$$\{1\} = G_0 \subset G_1 \subset \cdots \subset G_n = \mathscr{G}(F_n/F_0)$$

is a normal series. The factors of this series are the groups

$$\frac{G_i}{G_{i-1}} = \frac{\mathscr{G}(F_n/F_{n-i})}{\mathscr{G}(F_n/F_{n-i+1})} \approx \mathscr{G}(F_{n-i+1}/F_{n-i})$$

by **130**, and by **141**, $\mathscr{G}(F_{n-i+i}/F_{n-i})$ is solvable. Thus, $\mathscr{G}(F_n/F_0)$ has a normal series with solvable factors and therefore is a solvable group itself (**75β**). Applying the fundamental theorem (**130**) to the fields $F \subset E \subset F_n$, we have

$$\mathscr{G}(E/F) \approx \frac{\mathscr{G}(F_n/F)}{\mathscr{G}(F_n/E)}.$$

Since $\mathscr{G}(E/F)$ is isomorphic to a quotient group of the solvable group $\mathscr{G}(F_n/F)$, it follows from **75** that $\mathscr{G}(E/F)$ is solvable.

On the other hand, suppose that $\mathscr{G}(E/F)$ is a solvable group. Let

$$\{1\} = G_0 \subset G_1 \subset \cdots \subset G_n = \mathscr{G}(E/F)$$

be a composition series for $\mathscr{G}(E/F)$. Let F_i denote the fixed field of the group G_{n-i}. Then we have a tower over F,

$$F = F_0 \subset F_1 \subset \cdots \subset F_n = E.$$

Furthermore, $\mathscr{G}(E/F_i) = G_{n-i}$. Now G_{n-i} is a normal subgroup of G_{n-i+1}, and by the fundamental theorem (**130**) we may conclude that F_i is a Galois extension of F_{i-1} with

$$\mathscr{G}(F_i/F_{i-1}) \approx \frac{\mathscr{G}(E/F_{i-1})}{\mathscr{G}(E/F_i)} = \frac{G_{n-i+1}}{G_{n-i}}.$$

Since $\mathscr{G}(E/F)$ is solvable, the group G_{n-i+1}/G_{n-i} is cyclic (of prime order), and therefore $\mathscr{G}(F_i/F_{i-1})$ is cyclic. By **144** F_i is a radical extension of F_{i-1}. Applying **143** inductively yields in a finite number of steps that $E = F_n$ is a radical extension of $F_0 = F$, and the proof is complete.

145α. Let $fx = 0$ be an equation of degree 6 which is solvable by radicals. Prove that $fx = 0$ is solvable by the extraction of square roots, cube roots, and fifth roots only.

145β. Prove that equations of degree 2, 3, and 4 must be solvable by radicals.

146. *Quadratic Equations.* The simplest possible example of the preceding theory is the solution of the quadratic equation

$$x^2 - px + q = 0.$$

We assume that p and q are elements of a number field F and that $x^2 - px + q$ is irreducible over F with splitting field E. As a result $[E:F] = 2$, and $\mathcal{G}(E/F) = \{1, \phi\}$ is a cyclic group of order 2. Furthermore, F contains the square roots of unity, ± 1. If α_1 and α_2 are the roots of $x^2 - px + q$, then $p = \alpha_1 + \alpha_2$ and $q = \alpha_1\alpha_2$. According to **144**, we can solve the equation by means of the Lagrange resolvents. We compute:

$$(1, \alpha_1) = \alpha_1 + \phi\alpha_1 = \alpha_1 + \alpha_2 = p,$$
$$(-1, \alpha_1) = \alpha_1 - \phi\alpha_1 = \alpha_1 - \alpha_2,$$
$$(1, \alpha_2) = \alpha_2 + \phi\alpha_2 = \alpha_2 + \alpha_1 = p,$$
$$(-1, \alpha_2) = \alpha_2 - \phi\alpha_2 = \alpha_2 - \alpha_1 = -(-1, \alpha_1).$$

If we let $\xi = (-1, \alpha_1) = -(-1, \alpha_2)$, then we have

$$\alpha_1 = \tfrac{1}{2}\{(1, \alpha_1) + (-1, \alpha_2)\} = \tfrac{1}{2}(p + \xi),$$
$$\alpha_2 = \tfrac{1}{2}\{(1, \alpha_2) + (-1, \alpha_2)\} = \tfrac{1}{2}(p - \xi).$$

Now the theory predicts that the squares of the LaGrange resolvents will be elements of F. This is obviously true for $(1, \alpha_1)$ and $(1, \alpha_2)$ because $p \in F$. However, we also have

$$\xi^2 = \alpha_1^2 - 2\alpha_1\alpha_2 + \alpha_2^2 = (\alpha_1 + \alpha_2)^2 - 4\alpha_1\alpha_2 = p^2 - 4q.$$

Consequently, $\xi = \pm\sqrt{p^2 - 4q}$, and finally we obtain

$$\alpha_1, \alpha_2 = \tfrac{1}{2}(p \pm \sqrt{p^2 - 4q}).$$

147. *Cubic Equations.* The first case of any complexity among the examples of the preceding theory is the cubic equation

$$x^3 - px^2 + qx - r = 0. \tag{1}$$

We assume that p, q, and r are elements of F, a number field containing the cube roots of unity 1, ρ, ρ^2. (Since we have that

$$\rho = -\tfrac{1}{2} + \tfrac{1}{2}\sqrt{-3} \quad \text{and} \quad \rho^2 = -\tfrac{1}{2} - \tfrac{1}{2}\sqrt{-3},$$

it is enough that F contain $\sqrt{-3}$.) Let E denote the splitting field of $x^3 - px^2 + qx - r = 0$. E is a Galois extension of F by **132**, and $\mathcal{G}(E/F)$ is a permutation group of the roots, α_1, α_2, α_3 of (1). For the sake of argument we shall suppose that $\mathcal{G}(E/F)$ has all the permutations of α_1, α_2, α_3, or in other words, that $\mathcal{G}(E/F) \approx S_3$ and $[E: F] = o(S_3) = 6$. Of course S_3 is a solvable group: the composition series

$$\{e\} \subset A_3 \subset S_3$$

has cyclic factors A_3 of order 3 and S_3/A_3 of order 2. Let B denote the fixed field of A_3, that is, the subfield of E which remains fixed under all even permutations of $\alpha_1, \alpha_2, \alpha_3$. Now A_3 is a normal subgroup of S_3, and consequently, B is a Galois extension of F with $\mathcal{G}(B/F) \approx S_3/A_3$ and $[B: F] = 2$. Clearly,

$$\Delta = (\alpha_1 - \alpha_2)(\alpha_2 - \alpha_3)(\alpha_3 - \alpha_1)$$

is an element of B. Since permutations of $\alpha_1, \alpha_2, \alpha_3$ carry Δ to $\pm\Delta$, they leave Δ^2 fixed and therefore $\Delta^2 \in F$. We can compute Δ^2 in terms of the elementary symmetric functions of $\alpha_1, \alpha_2, \alpha_3$ using the fact that

$$\alpha_1 + \alpha_2 + \alpha_3 = p,$$
$$\alpha_1\alpha_2 + \alpha_2\alpha_3 + \alpha_3\alpha_1 = q,$$
$$\alpha_1\alpha_2\alpha_3 = r.$$

In fact we have, after a lengthy computation (given below),

$$\Delta^2 = -4p^3r - 27r^2 + 18pqr - 4q^3 + p^2q^2. \tag{2}$$

Clearly, $B = F(\Delta)$ and every element of B can be written in the form $u + v\Delta$ where $u, v \in F$. Now E is a Galois extension of B with Galois group $\mathcal{G}(E/B) \approx A_3$, a cyclic group of order 3. The Lagrange resolvents for α_1 are given by

$$\begin{aligned}
(1, \alpha_1) &= \alpha_1 + \alpha_2 + \alpha_3 = p, \\
(\rho, \alpha_1) &= \alpha_1 + \rho\alpha_2 + \rho^2\alpha_3, \\
(\rho^2, \alpha_1) &= \alpha_1 + \rho^2\alpha_2 + \rho\alpha_3.
\end{aligned} \tag{3}$$

(We have assumed the choice of a generator for $\mathcal{G}(E/B)$ which cyclically permutes $\alpha_1, \alpha_2, \alpha_3$.) The cubes $(\rho, \alpha_1)^3$ and $(\rho^2, \alpha_1)^3$ are elements of B which we compute as follows:

$$\begin{aligned}
(\rho, \alpha_1)^3 &= \alpha_1^3 + \alpha_2^3 + \alpha_3^3 + 3\rho(\alpha_1^2\alpha_2 + \alpha_2^2\alpha_3 + \alpha_3^2\alpha_1) \\
&\quad + 3\rho^2(\alpha_1\alpha_2^2 + \alpha_2\alpha_3^2 + \alpha_3\alpha_1^2) + 6\alpha_1\alpha_2\alpha_3 \\
&= (\alpha_1 + \alpha_2 + \alpha_3)^3 + (3\rho - 3)(\alpha_1^2\alpha_2 + \alpha_2^2\alpha_3 + \alpha_3^2\alpha_1) \\
&\quad + (3\rho^2 - 3)(\alpha_1\alpha_2^2 + \alpha_2\alpha_3^2 + \alpha_3\alpha_1^2) \\
&= p^3 - \tfrac{9}{2}(\alpha_1^2\alpha_2 + \alpha_2^2\alpha_1 + \alpha_2^2\alpha_3 + \alpha_2\alpha_3^2 + \alpha_3^2\alpha_1 + \alpha_1\alpha_3^2) \\
&\quad + \tfrac{3}{2}\sqrt{-3}(\alpha_1^2\alpha_2 - \alpha_2^2\alpha_1 + \alpha_2^2\alpha_3 - \alpha_2\alpha_3^2 + \alpha_3^2\alpha_1 - \alpha_1\alpha_3^2) \\
(\rho, \alpha_1)^3 &= p^3 - \tfrac{9}{2}(pq - 3r) - \tfrac{3}{2}\sqrt{-3}\Delta.
\end{aligned} \tag{4}$$

Similarly, we may compute

$$(\rho^2, \alpha_1)^3 = p^3 - \tfrac{9}{2}(pq - 3r) + \tfrac{3}{2}\sqrt{-3\Delta}.$$

Next, we note that

$$(\rho, \alpha_1)(\rho^2, \alpha_1) = \alpha_1^2 + \alpha_2^2 + \alpha_3^2 + (\rho + \rho^2)(\alpha_1\alpha_2 + \alpha_2\alpha_3 + \alpha_3\alpha_1)$$
$$= p^2 - 3q.$$

Finally to write the solutions of (1), we let ξ_1 be any one of the three cube roots of

$$p^3 - \tfrac{9}{2}(pq - 3r) - \tfrac{3}{2}\sqrt{-3\Delta}$$

and determine ξ_2 by

$$\xi_1\xi_2 = p^2 - 3q.$$

If we set $(\rho, \alpha_1) = \xi_1$ and $(\rho^2, \alpha_1) = \xi_2$, then we have

$$(1, \alpha_3) = (1, \alpha_2) = (1, \alpha_1) = p,$$
$$(\rho, \alpha_2) = \rho^2(\rho, \alpha_1) = \rho^2\xi_1,$$
$$(\rho^2, \alpha_2) = \rho(\rho^2, \alpha_1) = \rho\xi_2,$$
$$(\rho, \alpha_3) = \rho(\rho, \alpha_1) = \rho\xi_1,$$
$$(\rho^2, \alpha_3) = \rho^2(\rho^2, \alpha_1) = \rho^2\xi_2,$$

and consequently,

$$\alpha_1 = \tfrac{1}{3}(p + \xi_1 + \xi_2),$$
$$\alpha_2 = \tfrac{1}{3}(p + \rho^2\xi_1 + \rho\xi_2), \tag{5}$$
$$\alpha_3 = \tfrac{1}{3}(p + \rho\xi_1 + \rho^2\xi_2).$$

Although our argument was motivated by the assumption that $\mathscr{G}(E/F) \approx S_3$, *all the computations involved are completely general, and therefore the equations (5) represent the solutions of the general cubic equations* (1).

Remarks. For the special case in which $p = 0$ in (1), the formulas of (5) become much simpler:

$$\alpha_1, \alpha_2, \alpha_3 = \sqrt[3]{\frac{r}{2} + \sqrt{\frac{r^2}{4} + \frac{q^3}{27}}} + \sqrt[3]{\frac{r}{2} - \sqrt{\frac{r^2}{4} + \frac{q^3}{27}}},$$

where the cube roots are varied and the product of the two terms is always $-q/3$ for any root. This equation is known as *Cardan's Formula*. The general case can always be reduced to this special one by the substitution $x = X + p/3$ in (1).

Computation of the Discriminant, Δ^2. The quantity

$$\Delta^2 = (\alpha_1 - \alpha_2)^2(\alpha_2 - \alpha_3)^2(\alpha_3 - \alpha_1)^2$$

is called the *discriminant* because it vanishes whenever two of the roots α_1, α_2, α_3 are equal. To derive formula (2) above, we observe that Δ can be expressed as *Vandermonde's determinant*:

$$\Delta = \det \begin{vmatrix} 1 & 1 & 1 \\ \alpha_1 & \alpha_2 & \alpha_3 \\ \alpha_1^2 & \alpha_2^2 & \alpha_3^2 \end{vmatrix} = \det \begin{vmatrix} 1 & \alpha_1 & \alpha_1^2 \\ 1 & \alpha_2 & \alpha_2^2 \\ 1 & \alpha_3 & \alpha_3^2 \end{vmatrix},$$

and therefore Δ^2 can be expressed as the determinant of the product of the two matrices. In other words,

$$\Delta^2 = \det \begin{vmatrix} \pi_0 & \pi_1 & \pi_2 \\ \pi_1 & \pi_2 & \pi_3 \\ \pi_2 & \pi_3 & \pi_4 \end{vmatrix}$$

$$= \pi_0 \pi_2 \pi_4 + 2\pi_1 \pi_2 \pi_3 - \pi_2^3 - \pi_0 \pi_3^2 - \pi_1^2 \pi_4,$$

where $\pi_i = \alpha_1^i + \alpha_2^i + \alpha_3^i$. Now we have (**131α** and **131β**),

$$\pi_0 = 1 + 1 + 1 = 3,$$
$$\pi_1 = \alpha_1 + \alpha_2 + \alpha_3 = p,$$
$$\pi_2 = \alpha_1^2 + \alpha_2^2 + \alpha_3^2 = p^2 - 2q,$$
$$\pi_3 = \alpha_1^3 + \alpha_2^3 + \alpha_3^3 = p^3 - 3pq + 3r,$$
$$\pi_4 = \alpha_1^4 + \alpha_2^4 + \alpha_3^4 = p^4 - 4p^2q + 4pr + 2q^2.$$

Substituting these values into the expression above for Δ^2 will yield equation (2).

147α.   Verify that the substitution $x = X + p/3$ in

$$x^3 - px^2 + qx - r = 0$$

yields an equation of the form

$$X^3 + QX - R = 0.$$

147β.   Derive Cardan's formulas from the solution of the cubic given by the formulas (5).

147γ. Use the method of **147** to solve the following cubic equations over **Q**:

$$x^3 - x^2 - x + 2 = 0,$$
$$x^3 - 6x^2 + 11x - 6 = 0,$$
$$x^3 + x + 3 = 0.$$

147δ. Devise cubic equations over **Q** whose Galois groups have orders 1, 2, 3, and 6.

147ε. Prove that a cubic equation over **Q** has three real roots if $\Delta^2 > 0$ and one real root if $\Delta^2 < 0$.

147ζ. Let f be a polynomial of degree 3 irreducible over **Q**. Prove that the splitting field of f is $\mathbf{Q}(\Delta, \alpha)$ where Δ^2 is the discriminant of f and α is one of its roots.

147η. Show that a cubic equation irreducible over **Q** with three real roots cannot be solved by real radicals alone.

148. *Quartic Equations.*

Let E denote the splitting field of the quartic equation

$$x^4 - px^3 + qx^2 - rx + s = 0 \tag{1}$$

over F, a number field containing the cube roots of unity. Just as with the quadratic and cubic equations, we shall assume that $\mathscr{G}(E/F)$ contains all permutations of the roots α_1, α_2, α_3, α_4 of (1), or in other words, that $\mathscr{G}(E/F) \approx S_4$ and $[E:F] = o(S_4) = 24$.

S_4 has A_4 as a normal subgroup, and A_4 has a normal subgroup N containing the identity e and (12)(34), (13)(24), (14)(23). N is abelian, and therefore the subgroup K containing e and (12)(34) is normal. Thus, S_4 has a composition series with cyclic factors:

$$\{e\} \subset K \subset N \subset A_4 \subset S_4.$$

We take B_1, B_2, and B_3 to be the fixed fields of A_4, N, and K, respectively.

As in the case of the cubic equation, $B_1 = F(\Delta)$, where

$$\Delta = (\alpha_1 - \alpha_2)(\alpha_1 - \alpha_3)(\alpha_1 - \alpha_4)(\alpha_2 - \alpha_3)(\alpha_2 - \alpha_4)(\alpha_3 - \alpha_4),$$

and $\Delta^2 \in F$. We shall not need to compute Δ^2—it will fall out in what follows.

The element $\theta_1 = \alpha_1\alpha_2 + \alpha_3\alpha_4$ is left fixed by N and by the permutations (12), (34), (1324), and (1423). Hence $\theta_1 \in B_2$. All the permutations of S_4 applied to θ_1 yield only the numbers

$$\theta_1 = \alpha_1\alpha_2 + \alpha_3\alpha_4,$$
$$\theta_2 = \alpha_1\alpha_3 + \alpha_2\alpha_4,$$
and
$$\theta_3 = \alpha_1\alpha_4 + \alpha_2\alpha_3.$$

Moreover, the elements of N are the only permutations leaving fixed all three numbers θ_1, θ_2, and θ_3. Consequently, $B_2 = F(\theta_1, \theta_2, \theta_3)$. (Why?) Furthermore the polynomial

$$(y - \theta_1)(y - \theta_2)(y - \theta_3) = y^3 - Py^2 + Qy - R \qquad (2)$$

is left fixed by S_4, and its coefficients belong to F. Thus, B_2 is the splitting field of (2) over F. (We call (2) the *resolvent cubic* of (1).) The coefficients P, Q, R may be computed as follows:

$$P = \theta_1 + \theta_2 + \theta_3 = \sigma_2(\alpha_1, \alpha_2, \alpha_3, \alpha_4) = q,$$
$$Q = \theta_1\theta_2 + \theta_2\theta_3 + \theta_3\theta_1 = pr - 4s,$$
$$R = \theta_1\theta_2\theta_3 = s(p^2 - 4q) + r^2.$$

(Details are left to the reader.) The discriminant of (2) is the quantity

$$(\theta_1 - \theta_2)^2(\theta_2 - \theta_3)^2(\theta_3 - \theta_1)^2$$
$$= (\alpha_2 - \alpha_3)^2(\alpha_1 - \alpha_4)^2(\alpha_3 - \alpha_4)^2(\alpha_1 - \alpha_2)^2(\alpha_1 - \alpha_3)^2(\alpha_2 - \alpha_4)^2, \qquad (3)$$

which is just the discriminant of Δ^2 of (1). The right-hand side of (3) may be computed from the formula of the preceding article. (We shall not need Δ for the solution of (1).) Of course the roots θ_1, θ_2, and θ_3 of (2) may be obtained from the formulas of **147**.

To complete the solution of (1), we set

$$\xi_1 = \alpha_1 + \alpha_2 - \alpha_3 - \alpha_4 = 2(\alpha_1 + \alpha_2) - p,$$
$$\xi_2 = \alpha_1 - \alpha_2 + \alpha_3 - \alpha_4 = 2(\alpha_1 + \alpha_3) - p,$$
$$\xi_3 = \alpha_1 - \alpha_2 - \alpha_3 + \alpha_4 = 2(\alpha_1 + \alpha_4) - p.$$

Since the permutations of N either leave fixed or change the sign of each ξ_i, it follows that $\xi_i^2 \in B_2$. Direct computation shows

$$\xi_1^2 = p^2 - 4q + 4\theta_1,$$
$$\xi_2^2 = p^2 - 4q + 4\theta_2, \qquad (4)$$
$$\xi_3^2 = p^2 - 4q + 4\theta_3.$$

Next, we note that

$$\xi_1\xi_2\xi_3 = 8(\alpha_1 + \alpha_2)(\alpha_1 + \alpha_3)(\alpha_1 + \alpha_4) - 4p(\alpha_1 + \alpha_2)(\alpha_1 + \alpha_3)$$
$$- 4p(\alpha_1 + \alpha_2)(\alpha_1 + \alpha_4) - 4p(\alpha_1 + \alpha_3)(\alpha_1 + \alpha_4)$$
$$+ 2p^2(3\alpha_1 + \alpha_2 + \alpha_3 + \alpha_4) - p^3$$

$$= 8\alpha_1^3 + 8(\alpha_2 + \alpha_3 + \alpha_4)\alpha_1^2 + 8(\alpha_2\alpha_3 + \alpha_3\alpha_4 + \alpha_4\alpha_2)\alpha_1 + 8\alpha_1\alpha_2\alpha_3$$
$$- 4p[3\alpha_1^2 + 2(\alpha_2 + \alpha_3 + \alpha_4)\alpha_1 + \alpha_2\alpha_3 + \alpha_3\alpha_4 + \alpha_4\alpha_2]$$
$$+ 2p^2(2\alpha_1 + p) - p^3$$

$$= 8p\alpha_1^2 + 8r - 4p[\alpha_1^2 + 2p\alpha_1 + q - \alpha_1(p - \alpha_1)]$$
$$+ 4p^2\alpha_1 + 2p^3 - p^3$$

$$= 8r - 4pq + p^3.$$

Finally, we write out the solutions of (1). We take ξ_1, ξ_2, ξ_3 to be the square roots indicated by (4), in such a way that $\xi_1\xi_2\xi_3 = 8r - 4pq + p^3$, and we set

$$\begin{aligned}
\alpha_1 &= \tfrac{1}{4}(p + \xi_1 + \xi_2 + \xi_3), \\
\alpha_2 &= \tfrac{1}{4}(p + \xi_1 - \xi_2 - \xi_3), \\
\alpha_3 &= \tfrac{1}{4}(p - \xi_1 + \xi_2 - \xi_3), \\
\alpha_4 &= \tfrac{1}{4}(p - \xi_1 - \xi_2 + \xi_3).
\end{aligned} \tag{5}$$

149. Quintic Equations. Unlike quadratic, cubic, or quartic equations, quintic equations are not in general solvable by radicals. To show this it is enough to give an example. We choose the polynomial

$$fx = 2x^5 - 10x + 5,$$

which is clearly irreducible over **Q** by the Eisenstein criterion (**107**). Let E denote the splitting field of f over **Q**. The Galois group $\mathscr{G}(E/\mathbf{Q})$ is a permutation group of the roots α_1, α_2, α_3, α_4, α_5 of f (**132**) and is therefore isomorphic to a subgroup of S_5. In fact, we shall show that $\mathscr{G}(E/\mathbf{Q}) \approx S_5$.

First we remark that $\mathscr{G}(E/\mathbf{Q})$ must be a transitive permutation group of the roots of f. In other words, given two roots α_i and α_j, there is some $\phi \in \mathscr{G}(E/\mathbf{Q})$ such that $\phi(\alpha_i) = \alpha_j$. If this were not the case, then the polynomial

$$gx = (x - \alpha_1)(x - \alpha') \cdots (x - \alpha''),$$

in which $\alpha_1, \alpha', \ldots, \alpha''$ are the distinct images of α_1 under $\mathscr{G}(E/\mathbf{Q})$, would be fixed by $\mathscr{G}(E/\mathbf{Q})$ and have coefficients in **Q**. What is more, g would be a proper divisor of f, contradicting irreducibility of f.

By elementary techniques of the calculus, we may sketch the graph of $\tfrac{1}{5}f$, which has the same roots as f. From Figure 16 we see that f has three real roots, which we call α_3, α_4, and α_5. The other roots, α_1 and α_2, must be complex and conjugate.

The automorphism of **C** which carries each complex number $a + bi$ to its complex conjugate, $a - bi$, simply interchanges α_1 and α_2 and fixes $\alpha_3, \alpha_4, \alpha_5$. Consequently, it restricts to an automorphism ϕ of $E = \mathbf{Q}(\alpha_1, \alpha_2, \alpha_3, \alpha_4, \alpha_5)$. Clearly, $\phi \in \mathscr{G}(E/\mathbf{Q})$.

Now we have that $\mathscr{G}(E/\mathbf{Q})$ is isomorphic to a subgroup H of S_5, which is transitive and which contains the transposition $(1, 2)$. By **86**, $H = S_5$ and

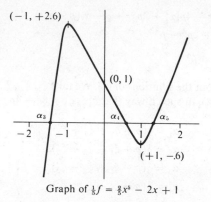

Graph of $\frac{1}{3}f = \frac{2}{3}x^5 - 2x + 1$

Figure 16

$\mathcal{G}(E/\mathbf{Q}) \approx S_5$. By **84** it follows that $\mathcal{G}(E/\mathbf{Q})$ is not solvable. Consequently, the equation

$$2x^5 - 10x + 5 = 0$$

is not solvable by radicals according to **145**.

149α.   Construct a polynomial of degree 7 which is irreducible over **Q** and not solvable by radicals.

149β.   Prove that for any prime $p > 3$ there exists a polynomial f of degree p which is irreducible over **Q** and not solvable by radicals.

Ring Theory

Chapter 5

A ring is an algebraic structure with two operations, addition and multiplication, but without all the properties required of those operations in field structure. Specifically, it is not required that every nonzero element of a ring have a multiplicative inverse. If we think of field structure as an abstraction of the properties of the set of rational numbers, then we should think of ring structure as an abstraction of the properties of the set of integers.

In this chapter we present the elementary theory of rings for commutative rings with unity. The main aim of this presentation is the proper abstract setting for unique factorization theorems like those for natural numbers (**24**) and polynomials (**104**). To show that this effort is worthwhile, the theory is applied to a special case of Fermat's last theorem (**175**).

Definition and
Examples of Ring Structure

150. A *ring* is an additive abelian group with an operation (written multiplicatively and called the *ring product*) which assigns to each ordered pair (a, b) of elements of R an element ab of R in such a way that:

(1) *multiplication is distributive over addition*; that is, for any three elements $a, b, c \in R$,

$$a(b + c) = ab + ac \quad \text{and} \quad (a + b)c = ac + bc;$$

(2) *multiplication is associative*; that is, for any three elements $a, b, c \in R$,

$$a(bc) = (ab)c;$$

(3) *multiplication is commutative*; that is, for any two elements $a, b \in R$,

$$ab = ba;$$

(4) *there exists a unity element* $1 \in R$ such that $1a = a = a1$ for every element $a \in R$.

Remarks. It is customary to require only conditions (1) and (2) in the definition of ring structure. In this case, an object satisfying all the conditions of the definition above is called a *commutative ring with unity*. Our investigation of ring theory will be confined to such objects, and use of the definition above avoids tiresome repetition of the phrase "commutative ring with unity."

150α. Indicate which of the following sets are rings. Unless otherwise specified, addition and multiplication are to be interpreted in the usual sense. For those which are not rings, specify which property of ring structure fails to hold.

(a) The set of integers, **Z**.
(b) The set of even integers, $2\mathbf{Z}$.
(c) The set of congruence classes mod n, $\mathbf{Z}_n$.
(d) The set of rational numbers, **Q**.
(e) The set of positive rational numbers, $\mathbf{Q}^+$.
(f) The set of real numbers, **R**.
(g) The set of complex numbers, **C**.
(h) The set of imaginary numbers.
 (i) The set $\mathbf{Z}(\sqrt{-3})$ of numbers $a + b\sqrt{-3}$ where $a, b \in \mathbf{Z}$.
 (j) The set $F[x]$ of polynomials in x over a field F.
(k) The set of polynomials over **Z**.
 (l) The set of primitive polynomials over **Z** (**105**).
(m) The set of all 2×2 matrices with real entries.
(n) The set of all continuous functions from **R** to **R**.
(o) The set of all power series with real coefficients.
(p) The set of all rational numbers with denominators not divisible by a given prime.
(q) The power set 2^X (**14**) of a set X, with union as addition and intersection as multiplication.

(r) The power set 2^X of a set X, with the symmetric difference as addition and intersection as multiplication.

(s) The set $\mathscr{E}(G)$ of endomorphisms (**60**) of an abelian group with addition defined by $(\phi_1 + \phi_2)g = (\phi_1 g) + (\phi_2 g)$, and with composition as multiplication.

(t) The set of integers **Z** with addition $\oplus$ and multiplication $\otimes$ defined by

$$a \oplus b = a + b + 1 \quad \text{and} \quad a \otimes b = ab + a + b.$$

150β. An algebraic structure satisfying the definition of ring structure except for commutativity of the product (statement (3) of **150**) will be called a *noncommutative ring*. If R is a noncommutative ring, let a new multiplication on R be given by $a * b = ab + ba$ for every pair of elements $a, b \in R$. When is R with this new multiplication a ring as defined in **150**?

150γ. An algebraic structure satisfying the definition of ring structure except for the existence of unity (statement (4) of **150**) will be called *a ring without unity*. If R is a ring without unity, define addition and multiplication on the set $\mathbf{Z} \times R$ by

$$(m, a) + (n, b) = (m + n, a + b)$$

and

$$(m, a)(n, b) = (mn, mb + na + ab).$$

Show that $\mathbf{Z} \times R$ is a ring with these operations and has a unity.

151. The additive identity element of a ring is called the *zero* (*element*) and denoted 0. The multiplicative identity element is called the *unity* (*element*) and denoted 1. The additive inverse of an element a is written $-a$. Clearly, $0a = 0$ and $(-1)a = -a$ for any ring element a.

A set with a single element has a unique addition and multiplication under which it is a ring. Such a ring is called *trivial* or *null*. We write $R = 0$ to indicate that a ring R is trivial. In a trivial ring $1 = 0$, which is to say, unity and zero coincide. A nontrivial ring must contain some nonzero element a, and since $1a = a \neq 0 = 0a$, we conclude that $1 \neq 0$. To summarize: *a ring is trivial if and only if unity is zero.*

In a nontrivial ring a nonzero element a may have a multiplicative inverse, that is, there may exist an element a^{-1} such that $aa^{-1} = 1 = a^{-1}a$. Such an element is called a *unit* of the ring. Clearly, the set of units in a ring forms a group under the ring product, and this group is called the *group of units* of the ring. A nonzero element which has no multiplicative inverse will be called a *proper element*. Thus, the elements of any ring are divided into three classes: zero, units, and proper elements.

shall call $\mathbf{Z}(\zeta)$ a *Kummer ring* after the mathematician E. E. Kummer (1810–1893), who studied the problem of unique factorization for these domains.

156.   *Polynomial Rings*.   For any ring R the set of polynomials in the variable x with coefficients in R is a ring under the usual addition and multiplication of polynomials. We denote this ring by $R[x]$. To give a precise status to $R[x]$, we shall adopt the following more formal definition.

Let $\tilde{\mathbf{N}}$ denote the set of nonnegative integers $\{0, 1, 2, \ldots\}$. A *polynomial f over the ring R* is a mapping $f: \tilde{\mathbf{N}} \to R$ which has the value 0 for all but a finite number of elements of $\tilde{\mathbf{N}}$. We let f_k denote the value of f on $k \in \tilde{\mathbf{N}}$. (Of course we are secretly thinking of f as $f_0 + f_1 x + f_2 x^2 + \cdots + f_n x^n$ with $f_k = 0$ for $k > n$.) Now $R[x]$ denotes the set of all such polynomials over R. Addition and multiplication of elements of $R[x]$ are defined by

$$(f + g)_k = f_k + g_k \quad \text{and} \quad (fg)_k = \sum_{i=0}^{k} f_i g_{k-i}.$$

If $f \in R[x]$ and $f_k = 0$ for all $k \in \tilde{\mathbf{N}}$, we write $f = 0$. If $f \neq 0$, then we may define the *degree* of f by

$$\deg f = \max\{k \in \tilde{\mathbf{N}} \mid f_k \neq 0\}.$$

Finally, we observe that we may identify R with a subring of $R[x]$ by letting $a \in R$ correspond to $f \in R[x]$, where $f_0 = a$ and $f_k = 0$ for $k > 0$.

This definition of $R[x]$ has the advantage that elements of $R[x]$ are defined as genuine mathematical entities and not as "expressions of the form . . .". It also has the advantage of allowing explicit definitions of addition and multiplication. Moreover, it generalizes easily to polynomials in several variables: a polynomial in n variables over R is simply a mapping

$$f: \tilde{\mathbf{N}} \times \cdots (n) \cdots \times \tilde{\mathbf{N}} \to R$$

which has the value 0 on all but a finite number of elements of its domain.

On the other hand, this formal definition of $R[x]$ has no real relation to the variable x. Another difficulty is that from force of habit we simply do not imagine a polynomial over R as a mapping from $\tilde{\mathbf{N}}$ to R.

Regardless of the manner in which $R[x]$ is defined, we may define the polynomial ring over R in two or more variables inductively by $R[x_1, x_2] = (R[x_1])[x_2]$, and so forth.

Proposition.   *If R is an integral domain, then $R[x]$ is also an integral domain.*

Proof.   If $f, g \in R[x]$ and $fg = 0$, then $f_0 g_0 = 0$ in R. Therefore, since R is an integral domain, either $f_0 = 0$ or $g_0 = 0$ (or both). Suppose that $f_0 = 0$. Then it follows from

$$(fg)_1 = f_1 g_0 + f_0 g_1 = f_1 g_0 = 0$$

that either $f_1 = 0$ or $g_0 = 0$. Continuing in this manner, we are forced to the conclusion that either $f = 0$ or $g = 0$.

Alternatively, we might suppose that f_i and g_j are the first nonzero coefficients of the polynomials f and g. Then $(fg)_{i+j} = f_i g_j \neq 0$, which shows that $fg \neq 0$. Thus the product of nonzero polynomials in $R[x]$ is nonzero.

Corollary. *If R is an integral domain, then $R[x_1, x_2, \ldots, x_n]$ is also an integral domain.*

156α. Determine the group of units of $R[x]$.

156β. What conditions on R will insure that $\deg(fg) = \deg f + \deg g$ for any two polynomials $f, g \in R[x]$?

156γ. The *ring of power series* $R[[x]]$ over a ring R is the set of *all* mappings $f: \tilde{N} \to R$ with the same rules of addition and multiplication as given above for $R[x]$. Determine the group of units of $R[[x]]$.

eals

---

157. An *ideal* of a ring R is an additive subgroup $\mathfrak{a}$ of R with the property that $r \in R$ and $a \in \mathfrak{a}$ imply $ra \in \mathfrak{a}$. Clearly, the set containing the single element 0 and the set consisting of the whole ring R are ideals. An ideal $\mathfrak{a}$ is called *proper* if $\mathfrak{a} \neq \{0\}$ and $\mathfrak{a} \neq R$.

157α. Prove that the intersection $\mathfrak{a} \cap \mathfrak{b}$ of two ideals $\mathfrak{a}$ and $\mathfrak{b}$ of a ring R is again an ideal of R.

157β. Prove that an ideal containing a unit element is the whole ring.

158. For each element a of a ring R, the set

$$(a) = \{x \in R \mid x = ra, r \in R\}$$

is an ideal, called the *principal ideal generated by a*. It is easy to see that the principal ideal (a) is the smallest ideal containing a. In other words, if $\mathfrak{a}$ is an ideal of R and $a \in \mathfrak{a}$, then $(a) \subset \mathfrak{a}$. We note that $(1) = R$ and consequently, $1 \in \mathfrak{a}$ implies $R \subset \mathfrak{a}$, or what is the same thing, $\mathfrak{a} = R$. An element $a \in R$ is clearly proper if and only if (a) is a proper ideal of R.

The integral domains in which every ideal is a principal ideal are of exceptional importance in ring theory. For brevity we refer to these integral domains as *principal ideal domains*. A field F can have only the improper ideals (0) and $(1) = F$. (Why?) Therefore a field is automatically a principal ideal domain. Fields, however, are the least interesting examples of principal ideal domains. The primary example of such a domain is Z, the ring of integers. To see this, we recall that every additive subgroup of Z has the form mZ (36) and observe that mZ is just the principal ideal (m).

158α.   Show that an element a of a ring R is a unit if and only if $(a) = R$.

158β.   Show that $(a) \subset (b)$ if and only if $a = rb$ for some r.

158γ.   Show that $(a) = (b)$ if and only if $a = ub$ for some unit element u.

158δ.   Prove that in the ring of integers Z, $(m) \cap (n) = ([m, n])$, where $[m, n]$ denotes the least common multiple of m and n (23γ).

158ε.   Let a and b be elements of a ring R. Show that the set

$$\mathfrak{c} = \{x \in R \mid x = ra + sb; r, s \in R\}$$

is an ideal of R and that it is the smallest ideal of R containing (a) and (b).

158ζ.   Let a and b be elements of a domain R. Show that the set

$$\mathfrak{c} = \{x \in R \mid ax \in (b)\}$$

is an ideal of R.

158η.   Prove that every ideal of the ring Z_n is principal. Is Z_n a principal ideal domain?

158θ.   Show that for $n > 1$ the ring of polynomials $R[x_1, x_2, \ldots, x_n]$ over a domain R is not a principal ideal domain.

159.   A *euclidean domain* is a nontrivial integral domain R together with a function, called the norm, $\delta: R^* \to N$ (where R^* denotes $R - \{0\}$), such that

(1) for all $a, b \in R^*$, $\delta(ab) = (\delta a)(\delta b)$,
(2) for all $a, b \in R^*$ there exist elements $q, r \in R$ such that $a = qb + r$ where $\delta r < \delta b$ or $r = 0$.

It clearly follows from **21** that the ring of integers Z is a euclidean domain setting $\delta a = |a|$ (absolute value). From **99** we see that the ring of polynomials $F[x]$ over a field F is a euclidean domain with $\delta f = 2^{\deg f}$ for $f \neq 0$. Now we give a fresh example.

Proposition. *The ring of Gaussian integers* $\mathbf{Z}(i)$ *together with the function* $\delta: \mathbf{Z}(i)^* \to \mathbf{N}$ *defined by* $\delta(u + vi) = u^2 + v^2$ *is a euclidean domain.*

Proof. First, we observe that

$$\delta(u + vi) = u^2 + v^2 = (u + vi)(u - vi) = |u + vi|^2.$$

Consequently, for $\alpha, \beta \in \mathbf{Z}(i)$, we have

$$\delta(\alpha\beta) = |\alpha\beta|^2 = |\alpha|^2 |\beta|^2 = (\delta\alpha)(\delta\beta).$$

Thus, (1) holds. To see that (2) holds, we use a little trick. $\mathbf{Z}(i)$ is a subring of the field $\mathbf{Q}(i)$. Thus, $\alpha, \beta \in \mathbf{Z}(i)$ implies $\alpha/\beta \in \mathbf{Q}(i)$, that is, $\alpha/\beta = u + vi$ for $u, v \in \mathbf{Q}$. Let p be the integer nearest u, and q the integer nearest v. Let $\gamma = p + qi \in \mathbf{Z}(i)$. Now we have $\alpha = \gamma\beta + \rho$ where

$$\delta\rho = |\beta((u - p) + (v - q)i)|^2 = |\beta|^2 ((u - p)^2 + (v - q)^2) \le \left(\frac{1}{2}\right) \delta\beta,$$

since $|u - p| \le 1/2$ and $|v - q| \le 1/2$.

Proposition. *A euclidean domain is a principal ideal domain.*

Proof. Let $\mathfrak{b}$ be a proper ideal of a euclidean domain R. Then among all the elements of $\mathfrak{b} \cap R^*$ there is (at least) one, say b, for which δb is a minimum. If $a \in \mathfrak{b} \cap R^*$, then $a = qb + r$ for some $q, r \in R$. Since $r = a - qb \in \mathfrak{b}$, we cannot have $\delta r < \delta b$ due to the choice of b. Therefore $r = 0$ and $b \mid a$ for any $a \in \mathfrak{b}$, or in other words, $\mathfrak{b} \subset (b)$. However, $(b) \subset \mathfrak{b}$ since $b \in \mathfrak{b}$, and as a result, $\mathfrak{b} = (b)$. This shows that R is a principal ideal domain.

Corollary. *The ring of polynomials $F[x]$ over a field F is a principal ideal domain.*

159α. Show that the ring $\mathbf{Z}(\omega)$, where

$$\omega = e^{2\pi i/3} = -\tfrac{1}{2} + \tfrac{1}{2}\sqrt{-3}$$

is a euclidean domain with $\delta(a + b\omega) = a^2 - ab + b^2$.

159β. Let R be a euclidean domain. We shall say that $d \in R$ is a greatest common divisor of the elements $a, b \in R$ (not both zero) if $c \mid a$ and $c \mid b$ imply that $c \mid d$ for any $c \in R$. Show that any pair of elements $a, b \in R$ (not both zero) must have a greatest common divisor d, which can be written as $ra + sb$ for some $r, s \in R$. (We interpret divisor in the usual sense: $r \mid t$ if and only if $r = st$ for some s.)

159γ. Let $\mathbf{Z}(\sqrt{-3})$ denote the ring of complex numbers of the form $a + b\sqrt{-3}$ where $a, b \in \mathbf{Z}$. Let $\delta(a + b\sqrt{-3}) = a^2 + 3b^2$. Is $\mathbf{Z}(\sqrt{-3})$ a euclidean domain? Is it a principal ideal domain?

159δ. Show that an element u of a euclidean domain is a unit if and only if $\delta(u) = 1$.

160. The *sum* of two ideals $\mathfrak{a}$ and $\mathfrak{b}$ of a ring R is the ideal

$$\mathfrak{a} + \mathfrak{b} = \{x \in R \mid x = a + b, a \in \mathfrak{a}, b \in \mathfrak{b}\}.$$

(It must be verified that the set $\mathfrak{a} + \mathfrak{b}$ defined above is an ideal of R, but this is routine.) Since every element $a \in \mathfrak{a}$ can be written as $a + 0$ and $0 \in \mathfrak{b}$, it follows that $\mathfrak{a} \subset \mathfrak{a} + \mathfrak{b}$. Similarly, $\mathfrak{b} \subset \mathfrak{a} + \mathfrak{b}$. In fact $\mathfrak{a} + \mathfrak{b}$ is the smallest ideal of R containing both $\mathfrak{a}$ and $\mathfrak{b}$, which is to say, $\mathfrak{a} \subset \mathfrak{c}$ and $\mathfrak{b} \subset \mathfrak{c}$ imply $\mathfrak{a} + \mathfrak{b} \subset \mathfrak{c}$.

160α. Prove the following properties of the sum of ideals:

(a) $(\mathfrak{a} + \mathfrak{b}) + \mathfrak{c} = \mathfrak{a} + (\mathfrak{b} + \mathfrak{c})$,
(b) $\mathfrak{a} + (0) = \mathfrak{a} = (0) + \mathfrak{a}$,
(c) $\mathfrak{a} + (1) = (1) = (1) + \mathfrak{a}$,
(d) $\mathfrak{a} + \mathfrak{b} = \mathfrak{b} + \mathfrak{a}$.

160β. Show that

$$\mathfrak{a} \cap (\mathfrak{b} + \mathfrak{c}) \supset (\mathfrak{a} \cap \mathfrak{b}) + (\mathfrak{a} \cap \mathfrak{c})$$

and

$$\mathfrak{a} + (\mathfrak{b} \cap \mathfrak{c}) \subset (\mathfrak{a} + \mathfrak{b}) \cap (\mathfrak{a} + \mathfrak{c})$$

for any three ideals $\mathfrak{a}, \mathfrak{b}, \mathfrak{c}$ of a ring R.

160γ. Let $(a_1, a_2, \ldots, a_n)$ denote the smallest ideal of a ring R containing the elements $a_1, a_2, \ldots, a_n \in R$. Prove that

$$(a_1, a_2, \ldots, a_n) = (a_1) + (a_2) + \cdots + (a_n).$$

160δ. Show that, in the ring of integers $\mathbf{Z}$, $(a) + (b) = (d)$ where d is the greatest common divisor of a and b.

161. The *product* $\mathfrak{a}\mathfrak{b}$ of two ideals $\mathfrak{a}$ and $\mathfrak{b}$ of a ring R is the smallest ideal of R containing all products of the form ab where $a \in \mathfrak{a}$ and $b \in \mathfrak{b}$. If $a_1, a_2, \ldots$, $a_n \in \mathfrak{a}$ and $b_1, b_2, \ldots, b_n \in \mathfrak{b}$, then $a_1 b_1, a_2 b_2, \ldots, a_n b_n \in \mathfrak{a}\mathfrak{b}$, and consequently, the sum

$$\sum_{i=1}^{n} a_i b_i = a_1 b_1 + a_2 b_2 + \cdots + a_n b_n \tag{1}$$

is an element of ab. In fact we could define ab as the set of all sums of the form (1). (Simple products a_1b_1 are included under the case $n = 1$.)

161α. Prove the following properties of the product of ideals:

- (a) $a(bc) = (ab)c$,
- (b) $a(1) = a = (1)a$,
- (c) $ab = ba$,
- (d) $ab \subset (a \cap b)$,
- (e) $(a)(b) = (ab)$.

161β. Prove that the product of ideals is distributive over the sum of ideals; that is,

$$a(b + c) = ab + ac$$

for any three ideals a, b, c of a ring R.

161γ. Show that $a(b \cap c) \subset ab \cap ac$ for ideals of a ring R.

161δ. Show that $a + b = R$ implies $a \cap b = ab$ for ideals a, b of R.

161ε. If a and b are ideals of a ring R, then we define their *quotient* to be the set

$$a : b = \{x \in R \mid xb \in a \text{ for all } b \in b\}.$$

Show that the quotient of two ideals is again an ideal of R.

161ζ. Show that the quotient operation on ideals in a ring R has the following properties:

- (a) $(a : b)b \subset a$,
- (b) $(a : b) : c = a : (bc)$,
- (c) $a : (b + c) = (a : b) \cap (a : c)$,
- (d) $(a \cap b) : c = (a : c) \cap (b : c)$,
- (e) $a : b = R$ if and only if $b \subset a$.

161η. In the ring of integers **Z** compute the ideals:

- (a) $(2) + (3)$,
- (b) $(2) + (4)$,
- (c) $(2) \cap ((3) + (4))$,
- (d) $(2)((3) \cap (4))$,
- (e) $(2)(3) \cap (2)(4)$,
- (f) $(6) \cap (8)$,
- (g) $(6)(8)$,
- (h) $(6) : (2)$,
- (i) $(2) : (6)$,
- (j) $(2) : (3)$.

165γ.　Prove that the characteristic of an integral domain is 0, 1, or a prime.

165δ.　If char $R_1 = n$ and char $R_2 = m$, what is char $(R_1 \times R_2)$? (See **152ε** for the definition of $R_1 \times R_2$.)

165ε.　Determine the number of ring homomorphisms $\mathbf{Z}_n \to \mathbf{Z}_m$.

166.　Proposition.　*If $\phi: R \to R'$ is a ring homomorphism and $\mathfrak{a}'$ is an ideal of R', then $\mathfrak{a} = \phi^{-1}\mathfrak{a}'$ is an ideal of R. Furthermore, if $\mathfrak{a}'$ is a prime ideal, then $\mathfrak{a}$ is also. If ϕ is an epimorphism and $\mathfrak{a}'$ is maximal, then $\mathfrak{a}$ is also maximal.*

Proof.　It is easy to see that $\mathfrak{a} = \phi^{-1}\mathfrak{a}'$ is an additive subgroup of R. If $a \in \mathfrak{a}$ and $r \in R$, then

$$\phi(ra) = \phi(r)\phi(a) \in \mathfrak{a}',$$

and thus $ra \in \mathfrak{a}$. This tells us that $\mathfrak{a}$ is an ideal of R. If $\mathfrak{a}'$ is prime and $ab \in \mathfrak{a}$, then

$$\phi(ab) = \phi(a)\phi(b) \in \mathfrak{a}'$$

and either

$$\phi a \in \mathfrak{a}' \quad \text{or} \quad \phi b \in \mathfrak{a}',$$

which proves either $a \in \mathfrak{a}$ or $b \in \mathfrak{a}$. Thus, $\mathfrak{a}$ is prime.

Suppose now that $\mathfrak{a}'$ is maximal, and that ϕ is onto. If $\mathfrak{a} \subset \mathfrak{b}$, then $\mathfrak{a}' \subset \phi\mathfrak{b}$. Since ϕ is an epimorphism, $\phi\mathfrak{b}$ is an ideal of R' as the reader will easily verify. Since $\mathfrak{a}'$ is maximal, we have either $\phi\mathfrak{b} = \mathfrak{a}'$, in which case $\mathfrak{b} \subset \phi^{-1}\phi\mathfrak{b} = \phi^{-1}\mathfrak{a}' = \mathfrak{a}$ and $\mathfrak{b} = \mathfrak{a}$, or else $\phi\mathfrak{b} = R'$. In this last case, $\mathfrak{b}$ contains an element b such that $\phi b = 1'$. Then $\phi(1 - b) = 0$ and $1 - b \in \mathfrak{a} \subset \mathfrak{b}$. Consequently, $1 = (1 - b) + b \in \mathfrak{b}$ and $\mathfrak{b} = R$. We have shown that $\mathfrak{a} \subset \mathfrak{b}$ implies $\mathfrak{b} = \mathfrak{a}$ or $\mathfrak{b} = R$, and therefore $\mathfrak{a}$ is maximal.

Remark.　In the last statement of the proposition we cannot remove the restriction that ϕ is an epimorphism. The inclusion mapping $i: \mathbf{Z} \to \mathbf{Q}$ gives an example: (0) is a maximal ideal of the field $\mathbf{Q}$, but $i^{-1}(0) = (0)$ is not maximal in $\mathbf{Z}$.

166α.　Prove that every ideal of the direct product $R_1 \times R_2$ of two rings has the form $\mathfrak{a}_1 \times \mathfrak{a}_2$ where $\mathfrak{a}_1$ and $\mathfrak{a}_2$ are ideals of R_1 and R_2, respectively.

166β.　Let ϕ be a ring epimorphism from R to R'. Show that

$$\phi^{-1}(\mathfrak{a}' + \mathfrak{b}') = (\phi^{-1}\mathfrak{a}') + (\phi^{-1}\mathfrak{b}')$$

and that

$$\phi^{-1}(\mathfrak{a}'\mathfrak{b}') = (\phi^{-1}\mathfrak{a}')(\phi^{-1}\mathfrak{b}')$$

where $\mathfrak{a}'$ and $\mathfrak{b}'$ are ideals of R'.

167.   Quotient Rings.   Let $\mathfrak{a}$ be an ideal of the ring R. We can define an equivalence relation on R by the rule: $a \equiv b \bmod \mathfrak{a}$ if and only if $a - b \in \mathfrak{a}$. This is called *congruence modulo the ideal* $\mathfrak{a}$. ($a \equiv b \bmod \mathfrak{a}$ is read "*a* is congruent to *b* modulo $\mathfrak{a}$.") Additively, $\mathfrak{a}$ is a normal subgroup of R, and congruence modulo $\mathfrak{a}$ is a special instance of congruence modulo a normal subgroup (**37**). This makes it unnecessary to verify that the properties (**17**) of an equivalence relation hold.

The equivalence class of $r \in R$ is denoted $r + \mathfrak{a}$; that is,

$$r + \mathfrak{a} = \{x \in R \mid x - r \in \mathfrak{a}\}$$
$$= \{x \subset R \mid x = r + a,\, a \in \mathfrak{a}\}.$$

The set of all equivalence classes of elements of R under congruence modulo $\mathfrak{a}$ is denoted $R/\mathfrak{a}$ since additively it is simply the quotient group of R by the normal subgroup $\mathfrak{a}$. It is clear that $R/\mathfrak{a}$ is an abelian group with addition defined by the rule

$$(a + \mathfrak{a}) + (b + \mathfrak{a}) = a + b + \mathfrak{a}.$$

Furthermore, $R/\mathfrak{a}$ is a ring in which multiplication is given by

$$(a + \mathfrak{a})(b + \mathfrak{a}) = ab + \mathfrak{a}.$$

The reader should demonstrate for his own satisfaction that this multiplication is well defined and that, furnished with these operations, $R/\mathfrak{a}$ satisfies the axioms of ring structure (**150**).

We note that if R is the ring of integers $\mathbf{Z}$, and $\mathfrak{a}$ is the principal ideal (n), then $R/\mathfrak{a} = \mathbf{Z}/(n)$ is the ring $\mathbf{Z}_n$ (**153**).

167α.   Let $\phi: R \to R'$ be a ring epimorphism. Prove that $R/(\mathrm{Ker}\ \phi)$ is isomorphic to R'. (Ker ϕ is defined in **165α.**)

167β.   Show that the set of ideals of the quotient ring $R/\mathfrak{a}$ is in one-to-one correspondence with the set of ideals of R containing $\mathfrak{a}$.

167γ.   Let $\mathfrak{a}$ and $\mathfrak{b}$ be ideals of a ring R such that $\mathfrak{a} \subset \mathfrak{b}$. Show that the mapping $\phi: R/\mathfrak{a} \to R/\mathfrak{b}$ given by $\phi(a + \mathfrak{a}) = a + \mathfrak{b}$ is a well-defined ring epimorphism, and compute Ker ϕ.

167δ.   Let $\mathfrak{a}$ denote the ideal of $\mathbf{Z}(i)$, consisting of all Gaussian integers $a + bi$ such that $a \equiv b \bmod 2$. Describe the quotient ring $\mathbf{Z}(i)/\mathfrak{a}$.

167ε.   Let $R^{[0,1]}$ denote the ring of continuous functions on the closed unit interval [0, 1]. Let $\mathfrak{a}$ denote the ideal of $R^{[0,1]}$ consisting of all continuous functions $f: [0, 1] \to R$ such that $f(1/2) = 0$. Describe the quotient ring $R^{[0,1]}/\mathfrak{a}$.

167ζ.   Let $\mathfrak{a}$ and $\mathfrak{b}$ be *relatively prime* ideals of a ring R, that is, $\mathfrak{a} + \mathfrak{b} = R$. Prove that

$$R/(\mathfrak{ab}) \approx (R/\mathfrak{a}) \times (R/\mathfrak{b}).$$

168.   Proposition.   *The quotient ring $R/\mathfrak{a}$ is an integral domain if and only if $\mathfrak{a}$ is a prime ideal. $R/\mathfrak{a}$ is a field if and only if $\mathfrak{a}$ is a maximal ideal.*

Proof.   The mapping $\phi: R \to R/\mathfrak{a}$ which assigns to each element $r \in R$ its equivalence class $\phi r = r + \mathfrak{a}$ in $R/\mathfrak{a}$ is a ring epimorphism. The ideal of $R/\mathfrak{a}$ containing only 0 is the equivalence class of $0 \in R$. In other words, in $R/\mathfrak{a}$, $(0) = \mathfrak{a}$, and further, $\phi^{-1}(0) = \mathfrak{a}$.

If $R/\mathfrak{a}$ is an integral domain, then (0) is a prime ideal of $R/\mathfrak{a}$ (**164**) and $\mathfrak{a} = \phi^{-1}(0)$ is a prime ideal of R (**166**). On the other hand, if $\mathfrak{a}$ is a prime ideal, $R/\mathfrak{a}$ cannot have zero divisors:

$$(a + \mathfrak{a})(b + \mathfrak{a}) = ab + \mathfrak{a} = \mathfrak{a}$$

implies $ab \in \mathfrak{a}$; hence, either $a \in \mathfrak{a}$ or $b \in \mathfrak{a}$, which implies $a + \mathfrak{a} = \mathfrak{a}$ or $b + \mathfrak{a} = \mathfrak{a}$.

Similarly, if $R/\mathfrak{a}$ is a field, then (0) is a maximal ideal of $R/\mathfrak{a}$ (**164**) and $\mathfrak{a} = \phi^{-1}(0)$ is a maximal ideal of R (**166**). On the other hand, suppose $\mathfrak{a}$ is maximal. An element $a + \mathfrak{a}$ is zero in $R/\mathfrak{a}$ if and only if $a \in \mathfrak{a}$. If $a \notin \mathfrak{a}$, then $(a) + \mathfrak{a} = R$ because $\mathfrak{a}$ is maximal. As a result, $1 \in (a) + \mathfrak{a}$, that is, $1 = a'a + a''$ for some $a' \in R$, $a'' \in \mathfrak{a}$. Now we have

$$(a' + \mathfrak{a})(a + \mathfrak{a}) = a'a + \mathfrak{a} = 1 + \mathfrak{a}.$$

Clearly $1 + \mathfrak{a}$ is the identity element of $R/\mathfrak{a}$ and $(a + \mathfrak{a})^{-1} = (a' + \mathfrak{a})$. Since every nonzero element of $R/\mathfrak{a}$ has an inverse, $R/\mathfrak{a}$ is a field.

Unique Factorization

169.   A factorization $r = r_1 r_2 \cdots r_k$ of an element r of a ring R is a *proper factorization* of r if each factor r_i is a proper element (not a unit or zero) of R. A factorization having units or zero among the factors is called *improper*.

Every element r of a ring has an improper factorization $r = 1r$, but not every element has a proper factorization. An element of a ring which has no proper factorization is called a *prime*. A proper element which is prime is called a *proper prime*. Clearly the unity element 1 is always a prime (improper), but the zero element 0 is a prime (improper) if and only if the ring is an integral domain. (Why?)

For example, the prime elements of **Z** are 0, ± 1, and $\pm p$, where $p \in$ **N** is prime in the ordinary sense (**22**). The primes in the polynomial ring $F[x]$ over a field F are the constant polynomials (improper) and the irreducible polynomials (proper).

Two elements of a ring are *associates* if each one is a multiple of the other by a unit. It is not difficult to see that association in this sense is an equivalence relation.

169α. Prove that two elements of an integral domain are associates if and only if they generate the same principal ideal.

169β. Prove that any associate of a prime is a prime.

169γ. Which of the numbers 3, 5, 7, 11, 13, 17, 19 are prime in the ring

$$\mathbf{Z}(\sqrt{2}) = \{x \in \mathbf{R} \mid x = a + b\sqrt{2}, a, b \in \mathbf{Z}\}?$$

170. If a and b are elements of a ring R, we say that a *divides* b (written $a \mid b$) if $b = ra$ for some $r \in R$. The set of all elements of R divisible by $a \in R$ is just the principal ideal (a). Furthermore, $a \mid b$ if and only if $(b) \subset (a)$.

An element $d \in R$ is a *greatest common divisor* of elements a and b of R provided

(a) $d \mid a$ and $d \mid b$,
(b) $c \in R$ and $c \mid a$, $c \mid b$ imply $c \mid d$.

Generally, there is not a unique greatest common divisor: if d is a greatest common divisor, then so is $d' = ud$, where u is a unit. In an integral domain R, any two greatest common divisors d and d' of a and b are associates. (We must have $d \mid d'$ and $d' \mid d$ so that $d' = rd$ and $d = r'd' = r'rd$ which implies $r'r = 1$ when $d \neq 0$; $d = 0$ implies $d' = rd = 0$.)

Proposition. *In a principal ideal domain R, an element d is a greatest common divisor of two elements a and b if and only if*

$$(d) = (a) + (b).$$

Proof. Since $d \mid a$ and $d \mid b$ imply $(a) \subset (d)$ and $(b) \subset (d)$, we have $(a) + (b) \subset (d)$ when d is *any* common divisor of a and b (**160**). Since R is a principal ideal domain, $(a) + (b) = (c)$ for some $c \in R$ and $(c) \subset (d)$, implies

$d \mid c$. On the other hand, $(a) \subset (c)$ and $(b) \subset (c)$ imply $c \mid a$ and $c \mid b$. Therefore, $c \mid d$ and $(d) \subset (c)$ if d is a greatest common divisor of a and b, and it follows that $(d) = (c) = (a) + (b)$.

Suppose $(d) = (a) + (b)$. Then $(a) \subset (d)$ and $(b) \subset (d)$, from which we conclude that $d \mid a$ and $d \mid b$. If $c \mid a$ and $c \mid b$, then as before $(a) \subset (c)$ and $(b) \subset (c)$ so that $(d) = (a) + (b) \subset (c)$, and therefore $c \mid d$. Thus, d is a greatest common divisor of a and b.

It is clear from this proposition that *in a principal ideal domain every pair of elements has a greatest common divisor*. The following corollary is another immediate consequence of this proposition.

Corollary.   *If an element d of a principal ideal domain R is a greatest common divisor of elements $a, b \in R$, then there exist elements $r, r' \in R$ such that $d = ra + r'b$.*

Corollary.   *If p is a prime element of a principal ideal domain R, then $p \mid ab$ implies $p \mid a$ or $p \mid b$.*

Proof.   Suppose $p \mid ab$ and $p \nmid a$. Then 1 is a greatest common divisor of p and a. (Why?) By the preceding corollary,

$$1 = rp + r'a$$

for some $r, r' \in R$. Then $b = brp + r'ab$ is divisible by p. (This is essentially the same proof as given in **23** and **103**.)

170α.   An element m of a ring R is a *least common multiple* of two elements $a, b \in R$ if and only if

(1) $a \mid m$ and $b \mid m$,
(2) $a \mid c$ and $b \mid c$ for any element $c \in R$, then $m \mid c$.

Show that m is a least common multiple of a and b in R if and only if $(m) = (a) \cap (b)$.

170β.   Let d be a greatest common divisor and m a least common multiple of elements a and b of a domain R. Show that dm and ab are associates.

171.   A *unique factorization ring* is a ring in which the following conditions hold:

(1) every proper element is a product of proper primes (not necessarily distinct);

(2) two factorizations of the same proper element as a product of proper primes have the same number of factors;

(3) if $r = p_1 p_2 \cdots p_k = q_1 q_2 \cdots q_k$ are two factorizations of a proper element r into a product of proper primes, then there exists a permutation of k letters, $\pi \in S_k$ such that p_i and $q_{\pi(i)}$ are associate primes for $i = 1, 2, \ldots, k$.

It is not possible to require more than this. Given a product of proper primes $p_1 p_2 \cdots p_k$, we can select units $u_1, u_2, \ldots, u_k$ so that their product $u_1 u_2 \cdots u_k = 1$, and then for any $\pi \in S_k$ we have $p_1 p_2 \cdots p_k = q_1 q_2 \cdots q_k$ where $q_{\pi(i)} = u_i p_i$. We note that even in the ring of integers we have twelve factorizations of 12:

$$12 = \begin{cases} 2 \cdot 2 \cdot 3 = 2 \cdot (-2)(-3) = (-2) \cdot 2 \cdot (-3) = (-2)(-2) \cdot 3, \\ 2 \cdot 3 \cdot 2 = 2 \cdot (-3)(-2) = (-2) \cdot 3 \cdot (-2) = (-2)(-3) \cdot 2, \\ 3 \cdot 2 \cdot 2 = 3 \cdot (-2)(-2) = (-3) \cdot 2 \cdot (-2) = (-3)(-2) \cdot 2. \end{cases}$$

We shall be mainly interested in *unique factorization domains*, that is to say, integral domains which satisfy (1), (2), and (3). We observe, however, that for p prime, $\mathbf{Z}_{p^n}$ is a unique factorization ring which is not an integral domain.

171α. Prove that $\mathbf{Z}_{p^n}$ is a unique factorization ring (p prime).

171β. Show that a quotient ring of a unique factorization ring need not be a unique factorization ring.

172. Theorem. *A euclidean domain is a unique factorization domain.*

Proof. First we remark that an element x in a euclidean domain R with norm $\delta: R^* \to \mathbf{N}$ is a unit if and only if $\delta(x) = 1$. Clearly $\delta(a) = \delta(1a) = \delta(1)\delta(a)$ implies $\delta(1) = 1$, from which it follows that a unit u has norm 1, because

$$1 = \delta(uu^{-1}) = \delta(u)\delta(u^{-1}).$$

Now suppose $\delta(x) = 1$. Then we have $1 = qx + r$ where $r = 0$, since $\delta(r) < \delta(x)$ is impossible. Therefore, $1 = qx$ and x is a unit.

Next we show by induction on norms that every element $r \in R$ with $\delta(r) > 1$ is a product of proper primes. Clearly, $\delta(r) = 2$ implies r is prime. Suppose that $1 < \delta(r) < n$ implies that r is a product of proper primes, and let $\delta(r) = n$. If r itself is a prime, then it is a proper prime and we are finished. If r is not a prime, then it has a proper factorization $r = ab$. Then $\delta(r) = \delta(a)\delta(b)$, from which we conclude that $1 < \delta(a) < n$ and $1 < \delta(b) < n$. By the induction

The kernel of ψ is easily seen to be $(x^2 + 1)$, the principal ideal of $Z_p[x]$ generated by $x^2 + 1$. Consequently, $Z_p(i)$ is isomorphic to the quotient ring $Z_p[x]/(x^2 + 1)$. We can summarize what we have proved so far in the statement:

$$Z(i)/(p) \approx Z_p(i) \approx Z_p[x]/(x^2 + 1).$$

From this we see that the eight statements below are equivalent:

(1) p is prime in $Z(i)$;
(2) (p) is a maximal prime ideal of $Z(i)$;
(3) $Z(i)/(p)$ is a field;
(4) $Z_p(i)$ is a field;
(5) $Z_p[x]/(x^2 + 1)$ is a field;
(6) $(x^2 + 1)$ is a maximal prime ideal of $Z_p[x]$;
(7) $x^2 + 1$ is irreducible over Z_p;
(8) $x^2 + 1$ has no root in Z_p.

Thus, the question of primeness of p in $Z(i)$ reduces to whether $x^2 + 1$ has a root in Z_p or not. In Z_2 we have $+1 = -1$ and $x^2 + 1$ has the root 1. If p is an odd prime, then a root α of $x^2 + 1$ over Z_p satisfies $\alpha^2 = -1$ and $\alpha^4 = 1$. Therefore α is an element of order 4 in the multiplicative group Z'_p of Z_p. Conversely, an element of order 4 in Z'_p is a root of $x^2 + 1$. (Why?) However, Z'_p is a cyclic group (**100**) and therefore has an element of order 4 if and only if 4 divides $o(Z'_p) = p - 1$. Thus, we see that $x^2 + 1$ has a root in Z_p if and only if $p = 2$ or $p \equiv 1 \bmod 4$. Finally, we have that p is prime in $Z(i)$ when $p \neq 2$ and $p \not\equiv 1 \bmod 4$, which is to say, when $p \equiv 3 \bmod 4$.

Corollary (Fermat). *Every prime p of the form $4m + 1$ can be written uniquely as the sum of two squares.*

Proof. Since p is prime in Z, but not in $Z(i)$, there is a Gaussian prime $a + bi$ with $a \neq 0 \neq b$, which divides p. Then $a - bi$ also divides p and $a^2 + b^2 = (a + bi)(a - bi)$ divides p^2. However, $a^2 + b^2$ is prime, therefore $a^2 + b^2 = p$. Uniqueness follows from unique factorization in $Z(i)$. (Why?)

Now we have completely determined all the primes of $Z(i)$. The improper primes are 0, ± 1, and $\pm i$. The proper primes are of the form $\pm p$, $\pm ip$, where $p \in N$ is a prime of the form $4m + 3$, and of the form $a + bi$ where $a^2 + b^2$ is prime.

174α. Determine all the primes of $Z(i)$ with absolute value 5 or less.

174β. Factor as a product of primes in $Z(i)$ the numbers 15 and $6 + 8i$.

174γ. Determine completely the natural numbers that can be written as the sum of two squares.

175. *Fermat's Last Theorem.* Pierre de Fermat (1601–1665), whom LaPlace characterized as "the true inventor of the differential calculus," discovered many theorems in number theory as well. His interest in the subject was aroused by the appearance in 1621 of Claude Bachet's edition of the *Arithmetica* of Diophantus. Beside the eighth proposition of the second book ("To divide a square number into two other square numbers"), Fermat in 1637 made the following scholium:

"To divide a cube into two cubes, or a fourth power into two fourth powers, and generally any power whatever beyond the second into two of the same denomination, is impossible. Of this fact I have discovered a very wonderful demonstration. This narrow margin would not take it."

This statement, that for $n > 2$ the equation $x^n + y^n = z^n$ has no solutions in which x, y, and z are natural numbers, is called *Fermat's last theorem* or *Fermat's great theorem* (as opposed to Fermat's "little" theorem given in **42**). Despite the strenuous efforts of many eminent mathematicians, among them Euler, Legendre, Abel, Gauss, Dirichlet, Cauchy, Kummer, Kronecker, and Hilbert, no general proof has been attained. It seems likely that Fermat was mistaken in believing he had a proof.

In the attempt to prove Fermat's last theorem, much valuable mathematics has developed. The classical ideal theory, which forms the subject of the next chapter, is one result. Our interest in Fermat's last theorem at this point is due to its connection with unique factorization in the Kummer rings $Z(\rho)$ defined in **155**.

For a long time it was thought that the ordinary laws of arithmetic, such as the division algorithm and unique factorization, must extend to the domains $Z(\rho)$, where $\rho = e^{2\pi i/p}$ and p is prime. Gabriel Lamé (1795–1870) gave a proof of Fermat's last theorem assuming unique factorization in $Z(\rho)$, in the year 1847. The error in Lamé's proof was observed by Joseph Liouville (1809–1882) and by Kummer. Cauchy, also in 1847, gave a false proof that $Z(\rho)$ is a euclidean domain. (The first prime for which this fails is 23.) In the years 1844 to 1851 Kummer developed a theory of unique factorization for these rings. Kummer's theory was eventually superseded by the theory of ideals developed by Dedekind and Kronecker, but he did succeed in proving Fermat's last theorem for a large class of exponents.

We shall consider here only the first case of Fermat's last theorem, the equation $x^3 + y^3 = z^3$. Evidently Fermat knew a proof of this case, but the first published proof is that of Euler's *Elements of Algebra*, Chapter XV, Volume II (second edition, 1774). A much simpler proof, due to Gauss, is based upon the arithmetic of $Z(\omega)$ where $\omega = e^{2\pi i/3}$. This is the proof we shall give.

$Z(\omega)$ is the set of all complex numbers of the form $a + b\omega$ where $a, b \in Z$. Note that $\omega^2 = \bar{\omega} = -1 - \omega$. Now $Z(\omega)$ is a euclidean domain (**159α**) and therefore a principal ideal domain and a unique factorization domain. The norm of $z = a + b\omega$ is

$$z\bar{z} = (a + b\omega)(a + b\omega^2) = a^2 - ab + b^2.$$

Since a unit must have norm 1, it follows that $\mathbf{Z}(\omega)$ has only the six units ± 1, $\pm\omega$, and $\pm\omega^2$. By arguments entirely similar to those for $\mathbf{Z}(i)$ in **174**, we obtain the following: *an element $a + b\omega \in \mathbf{Z}(\omega)$ with $a \neq 0 \neq b$ is a (proper) prime of $\mathbf{Z}(\omega)$ if and only if its norm $a^2 - ab + b^2$ is a (proper) prime of $\mathbf{Z}$.* The other primes of $\mathbf{Z}(\omega)$ have the form $\pm p$, $\pm p\omega$, or $\pm p\omega^2$ where $p \in \mathbf{Z}$ is a proper prime and $p \neq a^2 - ab + b^2$ for any integers a and b. These are characterized by the following result.

Theorem.   *A positive proper prime $p \in \mathbf{Z}$ is a proper prime of $\mathbf{Z}(\omega)$ if and only if $p \equiv 5 \bmod 6$ or $p = 2$.*

Proof.   We construct the ring $\mathbf{Z}_p(\omega)$ of elements $a + b\omega$ where $a, b \in \mathbf{Z}_p$ by defining addition and multiplication as follows:

$$(a + b\omega) + (c + d\omega) = (a + c) + (b + d)\omega$$

and

$$(a + b\omega)(c + d\omega) = (ac - bd) + (ad + bc - bd)\omega.$$

It follows (as in **174**) that we have ring isomorphisms:

$$\mathbf{Z}(\omega)/(p) \approx \mathbf{Z}_p(\omega) \approx \mathbf{Z}_p[x]/(x^2 + x + 1).$$

Thus the question of primeness of $p \in \mathbf{Z}(\omega)$ is equivalent to irreducibility of $x^2 + x + 1$ over $\mathbf{Z}_p$. Over $\mathbf{Z}_2$, $x^2 + x + 1$ is irreducible because neither element of $\mathbf{Z}_2$ can be a root. Now suppose p is an odd prime and $x^2 + x + 1$ has a root $\alpha \in \mathbf{Z}_p$. Clearly $\alpha \neq 0$ and $1/\alpha$ is also a root, since the product of the two roots of $x^2 + x + 1$ is 1. Since the sum of the roots is -1, we have $\alpha + (1/\alpha) = -1$. Squaring yields $\alpha^2 + 2 + (1/\alpha^2) = 1$, or $\alpha^4 + \alpha^2 + 1 = 0$, which shows that α^2 is also a root. The polynomial $x^2 + x + 1$ can have at most two roots in $\mathbf{Z}_p$, therefore $\alpha^2 = \alpha$ or $\alpha^2 = 1/\alpha$. If $\alpha^2 = \alpha$, then $\alpha = 1$ and $1 + 1 + 1 = 0$ in $\mathbf{Z}_p$, implying that $p = 3$. If $\alpha^2 = 1/\alpha$, then $\alpha^3 = 1$ and α is an element of order 3 in the cyclic group $\mathbf{Z}_p'$ of order $p - 1$. As a result, $3 \mid (p - 1)$. Since p is odd, $2 \mid (p - 1)$, and therefore $6 \mid (p - 1)$. To summarize, $x^2 + x + 1$ is reducible over $\mathbf{Z}_p$, or in other words $x^2 + x + 1$ has a root in $\mathbf{Z}_p$, precisely when $p = 3$ or $p \equiv 1 \bmod 6$. The theorem follows from the fact that p prime, $p \neq 3$, and $p \not\equiv 1 \bmod 6$ imply $p = 2$ or $p \equiv 5 \bmod 6$. (Why?)

Theorem.   *The equation $x^3 + y^3 = z^3$ has no solution in natural numbers.*

Proof.   We shall actually prove more: the equation $x^3 + y^3 = z^3$ has no solutions in $\mathbf{Z}(\omega)$ except the trivial ones in which x, y, or z is zero. Suppose

x, y, and z are three nonzero elements of $\mathbf{Z}(\omega)$ such that $x^3 + y^3 = z^3$. Clearly we may assume that x, y, and z are relatively prime in pairs, which is to say that in terms of ideals,

$$(x, y) = (y, z) = (z, x) = (1) = \mathbf{Z}(\omega). \tag{1}$$

Indeed, if two of the three quantities x, y, and z had a greatest common divisor d which was not a unit, then the whole equation would be divisible by d^3, and (x/d), (y/d), (z/d) would be a solution for which the assumption held.

We note that $1 - \omega$ has norm

$$(1 - \omega)(1 - \bar{\omega}) = (1 - \omega)(1 - \omega^2) = 3,$$

and is therefore prime in $\mathbf{Z}(\omega)$. Furthermore, its complex conjugate is

$$1 - \bar{\omega} = 1 - \omega^2 = \omega^3 - \omega^2 = \omega^2(\omega - 1) = -\omega^2(1 - \omega).$$

Since $-\omega^2$ is a unit, $1 - \omega$ and $1 - \bar{\omega}$ are associates. Therefore if $1 - \omega$ divides $\alpha \in \mathbf{Z}(\omega)$, then $1 - \omega$ also divides $\bar{\alpha}$, the complex conjugate of α. We observe that every element of $\mathbf{Z}(\omega)$ is congruent to 0, 1, or 2 modulo $(1 - \omega)$: given $a + b\omega \in \mathbf{Z}(\omega)$, we have $a + b = 3q + r$ where $0 \le r < 3$ and

$$a + b\omega \equiv a + b \equiv 3q + r \equiv r \bmod (1 - \omega).$$

(Of course, $2 \equiv -1 \bmod (1 - \omega)$ since $1 - \omega$ divides 3.)

Finally, we remark that $\alpha \equiv \pm 1 \bmod (1 - \omega)^\lambda$ implies that $\alpha^3 \equiv \pm 1 \bmod (1 - \omega)^{\lambda + 3}$. To see this, we write $\alpha = \pm 1 + \beta(1 - \omega)^\lambda$ and then

$$\begin{aligned}
\alpha^3 \mp 1 &= (\alpha \mp 1)(\alpha \mp \omega)(\alpha \mp \omega^2) \\
&= \beta(1 - \omega)^\lambda(\beta(1 - \omega)^\lambda + 1 \mp \omega)(\beta(1 - \omega)^\lambda \pm 1 \mp \omega^2) \\
&= (1 - \omega)^{\lambda + 2}\beta(\beta \pm 1)(\beta \pm (1 + \omega)).
\end{aligned}$$

One of the three quantities β, $\beta \pm 1$, and $\beta \pm (1 + \omega)$ must be divisible by $1 - \omega$. (Why?) Therefore $\alpha^3 \mp 1$ is divisible by $(1 - \omega)^{\lambda + 3}$.

The arithmetic of $\mathbf{Z}(\omega)$ is relevant to the equation $x^3 + y^3 = z^3$ precisely because within $\mathbf{Z}(\omega)$ we have the factorization

$$x^3 + y^3 = (x + y)(x\omega + y\omega^2)(x\omega^2 + y\omega). \tag{2}$$

The three factors $x + y$, $x\omega + y\omega^2$, and $x\omega^2 + y\omega$ are all congruent $\bmod (1 - \omega)$ and their sum is zero since $1 + \omega + \omega^2 = 0$. In addition, taken in pairs they are either relatively prime or else have $1 - \omega$ as greatest common divisor as we see from the equations

Classical Ideal Theory

Chapter 6

In this chapter we study those integral domains which have a unique factorization theorem for ideals. Such rings are called Dedekind domains, and their study is called classical ideal theory. We define a Dedekind domain to be an integral domain whose ideals have a certain property (invertibility in the field of fractions) and then prove that this is equivalent to unique factorization of ideals. Finally, we apply this theory to prove that the ring of integers in a Galois extension of the rational field **Q** is a Dedekind domain. From this we draw the conclusion that the Kummer rings are Dedekind domains. This re-establishes a form of unique factorization for the rings associated with Fermat's last theorem, where the problem of unique factorization first became critical.

Fields of Fractions

---

176. Let R be a nontrivial integral domain. We shall construct a field $\mathbf{Q}_R$ containing R as a subring by adding to R all the fractions r/s where $r \in R$ and $s \in R^* = R - \{0\}$. Naturally $\mathbf{Q}_R$ is called the *field of fractions* of R. The elements of R are identified with the fractions having denominator 1. The general construction of the field of fractions $\mathbf{Q}_R$ out of R is an exact parallel of the construction of the field of rational numbers $\mathbf{Q}$ out of the ring of integers $\mathbf{Z}$. It is helpful to keep this in mind.

We define an equivalence relation on the set $R \times R^*$ by the rule $(r_1, s_1) \sim (r_2, s_2)$ if and only if $r_1 s_2 = r_2 s_1$. (The reader should verify that $\sim$ is an equivalence relation.) The equivalence class of (r, s) will be denoted r/s. Clearly $r_1/s_1 = r_2/s_2$ if and only if $r_1 s_2 = r_2 s_1$. We let $\mathbf{Q}_R$ denote the set of all such equivalence classes. Addition and multiplication in $\mathbf{Q}_R$ are defined by

$$\left(\frac{r_1}{s_1}\right) + \left(\frac{r_2}{s_2}\right) = \left(\frac{r_1 s_2 + r_2 s_1}{s_1 s_2}\right)$$

and

$$\left(\frac{r_1}{s_1}\right)\left(\frac{r_2}{s_2}\right) = \left(\frac{r_1 r_2}{s_1 s_2}\right).$$

It is necessary to verify that these operations are well defined, but we omit doing it. $\mathbf{Q}_R$ is a field under these operations. The additive identity element 0 may be represented as $0/s$ for any $s \in R^*$. The multiplicative identity element 1 may be represented as s/s for any $s \in R^*$. The additive inverse of r/s is $-r/s$ and, if $r \neq 0$, the multiplicative inverse is s/r.

The mapping $\phi: R \to \mathbf{Q}_R$ given by $\phi r = r/1$ is easily seen to be a ring monomorphism. This enables us to identify the element $r \in R$ with the element $r/1 \in \mathbf{Q}_R$ and to think of R as a subring of $\mathbf{Q}_R$.

We observe that for the ring of polynomials $F[x]$ over a field F, the field of fractions $\mathbf{Q}_{F[x]}$ is the field $F(x)$ of rational functions over F. (See **98α**.)

176α. Let R be a nontrivial integral domain and F a field. Show that a ring monomorphism $\phi: R \to F$ can be extended uniquely to a field monomorphism $\phi: \mathbf{Q}_R \to F$.

176β. Let S be a subset of a ring R such that (1) S contains no zero divisors of R (hence $0 \notin S$) and (2) $a, b \in S$ implies $ab \in S$. Construct a ring R_S of fractions r/s where the denominators s are elements of S, by imitating the field of fractions construction.

176γ. Show that any ring R is a subring of some ring with the property that every element is a unit or a zero divisor.

176δ. Show that the field of fractions of $\mathbf{Z}(\sqrt{-5})$ is $\mathbf{Q}(\sqrt{-5})$.

176ε. Show that the field of fractions of the Kummer ring $\mathbf{Z}(\rho)$ is $\mathbf{Q}(\rho)$. (See **155**.)

177. The field of fractions $\mathbf{Q}_R$ of a nontrivial integral domain R, being a field, has only two ideals, (0) and (1). We can, however, introduce a certain class of subsets of $\mathbf{Q}_R$, called *fractionary ideals* of R, whose formal behavior resembles that of ideals of R. Fractionary ideals are important in the study of Dedekind domains.

Let R be an integral domain with $1 \neq 0$, that is, nontrivial. A fractionary ideal of R is an additive subgroup $\mathfrak{a}$ of $\mathbf{Q}_R$ such that

(1) $a \in \mathfrak{a}$ implies $ra \in \mathfrak{a}$ for all $r \in R$,
(2) there is some $r \in R^*$ such that $ra \in R$ for all $a \in \mathfrak{a}$.

Every ordinary ideal of R is a fractionary ideal of R. For clarity and emphasis, we call an ordinary ideal of R an *integral ideal*. Obviously a fractionary ideal is integral if and only if it is a subset of R.

With every element $r/s \in \mathbf{Q}_R$ we may associate the *principal fractionary ideal* (r/s) defined by

$$\left(\frac{r}{s}\right) = \left\{ x \in \mathbf{Q}_R \,|\, x = \frac{r'r}{s}, r' \in R \right\}.$$

In other words, (r/s) contains all the multiples of r/s by elements of R.

The notions of sum and product of ideals may be extended to fractionary ideals in the obvious fashion. Thus, if $\mathfrak{a}$ and $\mathfrak{b}$ are fractionary ideals of R, then $\mathfrak{a} + \mathfrak{b}$ is the fractionary ideal containing all elements of the form $a + b$ where $a \in \mathfrak{a}$ and $b \in \mathfrak{b}$. Furthermore, $\mathfrak{ab}$ is the smallest fractionary ideal containing all products ab where $a \in \mathfrak{a}$ and $b \in \mathfrak{b}$. Of course if $\mathfrak{a}$ and $\mathfrak{b}$ are integral ideals, then so are $\mathfrak{a} + \mathfrak{b}$ and $\mathfrak{ab}$.

177α.   Which of the properties given in exercises **160α**, **160β**, **161α**, **161β**, **161γ**, and **161δ** hold for fractionary ideals?

177β.   Compute the following for fractionary ideals of $\mathbf{Z}$:

(a) $(1/2) + (2/3)$,
(b) $(1/2)(2/3)$,
(c) $(1/2)((2/3) + (3/4))$,
(d) $(1/2) \cap (2/3)$,
(e) $(1/2) \cap (2/3)(3/4)$.

178.   If $\mathfrak{a}$ and $\mathfrak{b}$ are fractionary ideals of R and $\mathfrak{b} \neq (0)$, then we define their *quotient* to be the set

$$\mathfrak{a}/\mathfrak{b} = \{ x \in \mathbf{Q}_R \,|\, (x)\mathfrak{b} \subset \mathfrak{a} \}.$$

To see that $\mathfrak{a}/\mathfrak{b}$ is again a fractionary ideal of R, we observe that it is clearly an additive subgroup of $\mathbf{Q}_R$ which is closed under multiplication by elements of R, and that it has the special property of fractionary ideals. Specifically, if $r \in R^*$ is an element such that $ra \in R$ for all $a \in \mathfrak{a}$ and b is a nonzero element of $\mathfrak{b}$, then $rb \in R^*$ is an element such that $(rb)x \in R$ for all $x \in \mathfrak{a}/\mathfrak{b}$. (Why?)

Note that even when $\mathfrak{a}$ and $\mathfrak{b}$ are integral ideals, their quotient $\mathfrak{a}/\mathfrak{b}$ need not be an integral ideal. Thus, the fractionary quotient $\mathfrak{a}/\mathfrak{b}$ is formally a new operation; it is related to the integral quotient $\mathfrak{a} : \mathfrak{b}$ by the equation $\mathfrak{a} : \mathfrak{b} = (\mathfrak{a}/\mathfrak{b}) \cap R$ when $\mathfrak{a}$ and $\mathfrak{b}$ are integral. (See **161ε**.)

Proposition.   *Every fractionary ideal is the quotient of an integral ideal by a principal integral ideal.*

Proof.   If $\mathfrak{a}$ is a fractionary ideal of R, there exists an element $r \in R^*$ such that $ra \in R$ for all $a \in \mathfrak{a}$, or equivalently, such that $(r)\mathfrak{a}$ is an integral ideal. We shall see that $\mathfrak{a} = (r)\mathfrak{a}/(r)$. If $x \in \mathfrak{a}$, then $(r)(x) \subset (r)\mathfrak{a}$, and consequently, $x \in (r)\mathfrak{a}/(r)$. On the other hand, if $x \in (r)\mathfrak{a}/(r)$, then $rx \in (r)\mathfrak{a}$, or equivalently, $rx = sra$ for some $s \in R$ and $a \in \mathfrak{a}$. Since R is an integral domain and $r \neq 0$, this implies $x = sa \in \mathfrak{a}$.

Corollary.   *If R is a principal ideal domain where $1 \neq 0$, then every fractionary ideal of R is principal.*

Proof.   By the proposition, we know that every fractionary ideal of R has the form $(r)/(s)$ for $r, s \in R$, $s \neq 0$. It suffices to observe that $(r)/(s) = (r/s)$.

178α.   Prove the following properties of the quotient operation for fractionary ideals:

(a) $(\mathfrak{a}/\mathfrak{b})\mathfrak{b} \subset \mathfrak{a}$,
(b) $(\mathfrak{a}/\mathfrak{b})/\mathfrak{c} = \mathfrak{a}/\mathfrak{b}\mathfrak{c}$,
(c) $\mathfrak{a}/(\mathfrak{b} + \mathfrak{c}) = (\mathfrak{a}/\mathfrak{b}) \cap (\mathfrak{a}/\mathfrak{c})$,
(d) $(\mathfrak{a} \cap \mathfrak{b})/\mathfrak{c} = (\mathfrak{a}/\mathfrak{c}) \cap (\mathfrak{b}/\mathfrak{c})$,
(e) $R \subset \mathfrak{a}/\mathfrak{b}$ if and only if $\mathfrak{b} \subset \mathfrak{a}$.

178β.   Compute the quotient $\mathfrak{a}/\mathfrak{b}$ of the fractionary ideals $\mathfrak{a}$ and $\mathfrak{b}$ of the ring R in the following cases:

(a) $R = \mathbf{Z}$, $\mathfrak{a} = (1/2)$, $\mathfrak{b} = (2/3)$,
(b) $R = \mathbf{Z}(i)$, $\mathfrak{a} = (1/1 + i)$, $\mathfrak{b} = (i/2)$,
(c) $R = \mathbf{Q}[x]$, $\mathfrak{a} = (x/x + 1)$, $\mathfrak{b} = (1/x)$.

179.   A fractionary ideal $\mathfrak{a}$ of a nontrivial integral domain R is *invertible* if there exists a fractionary ideal $\mathfrak{b}$ of R such that

$$\mathfrak{a}\mathfrak{b} = (1) = R.$$

If $\mathfrak{a}\mathfrak{b} = (1)$, then clearly $\mathfrak{b} \subset (1)/\mathfrak{a}$. On the other hand, $x \in (1)/\mathfrak{a}$ implies $x \in (x) = (x)\mathfrak{a}\mathfrak{b} \subset \mathfrak{b}$ since $(x)\mathfrak{a} \subset (1)$. Thus $\mathfrak{a}\mathfrak{b} = (1)$ implies $\mathfrak{b} = (1)/\mathfrak{a}$. It is convenient to write $\mathfrak{a}^{-1}$ in place of $(1)/\mathfrak{a}$ when $\mathfrak{a}$ is invertible.

We note that a nonzero principal fractionary ideal (a) is always invertible and that $(a)^{-1} = (a^{-1})$, or in other words, $(1)/(a) = (1/a)$. In general there are invertible fractionary ideals which are not principal, but we can prove the following.

Proposition.   *An invertible fractionary ideal is finitely generated.*

Proof.   Let $\mathfrak{a}$ be an invertible fractionary ideal of R. Since $1 \in (1) = \mathfrak{a}^{-1}\mathfrak{a}$, we have

$$1 = a'_1 a_1 + a'_2 a_2 + \cdots + a'_n a_n$$

for some elements $a'_1, a'_2, \ldots, a'_n \in \mathfrak{a}^{-1}$ and $a_1, a_2, \ldots, a_n \in \mathfrak{a}$. If $x \in \mathfrak{a}$, then the elements $r_1 = xa'_1, r_2 = xa'_2, \ldots, r_n = xa'_n$ belong to R and we have, multiplying the equation above by x,

$$x = r_1 a_1 + r_2 a_2 + \cdots + r_n a_n.$$

This shows that $a_1, a_2, \ldots, a_n$ generate $\mathfrak{a}$, that is,

$$\mathfrak{a} = (a_1, a_2, \ldots, a_n) = (a_1) + (a_2) + \cdots + (a_n).$$

179α.   Prove that if $\mathfrak{a}$ and $\mathfrak{b}$ are fractionary ideals and $\mathfrak{b}$ is invertible, then $\mathfrak{a}/\mathfrak{b} = \mathfrak{a}\mathfrak{b}^{-1}$.

179β.   Prove that if $\mathfrak{a}$ and $\mathfrak{b}$ are invertible fractionary ideals, then $\mathfrak{a} \subset \mathfrak{b}$ if and only if $\mathfrak{b}^{-1} \subset \mathfrak{a}^{-1}$.

179γ.   Prove that two fractionary ideals are both invertible if and only if their product is invertible.

179δ.   Let R be a nontrivial integral domain. Show that the invertible fractionary ideals of R form a group under the product of ideals.

179ε.   Suppose that $\mathfrak{a}, \mathfrak{b}, \mathfrak{a} + \mathfrak{b}$, and $\mathfrak{a} \cap \mathfrak{b}$ are invertible fractionary ideals. Show that

$$(\mathfrak{a} + \mathfrak{b})^{-1} = \mathfrak{a}^{-1} \cap \mathfrak{b}^{-1},$$
$$(\mathfrak{a} \cap \mathfrak{b})^{-1} = \mathfrak{a}^{-1} + \mathfrak{b}^{-1}.$$

179ζ.   Show that the ideal (x, y) of $\mathbf{Q}[x, y]$ is not invertible.

179η.   Compute the inverse of the ideal $(2, 1 + \sqrt{-5})$ in $\mathbf{Z}(\sqrt{-5})$.

180. A *Dedekind domain* is a nontrivial integral domain in which every nonzero fractional ideal is invertible. Every fractional ideal of a principal ideal domain is principal (**178**) and, if nonzero, invertible (**179**). Therefore *a nontrivial principal ideal domain is a Dedekind domain.*

Richard Dedekind (1831–1916) showed that Kummer's ideal elements, which seemed mysterious and artificial, could be viewed as subsets of a ring having special properties (closure under addition and closure under multiplication by ring elements). Dedekind called these sets ideals and showed that in many domains they possess a unique factorization theorem. We shall see that a nontrivial integral domain has unique factorization for ideals if and only if it is a Dedekind domain.

180α. Show that a nontrivial integral domain is a Dedekind domain if and only if for each integral ideal $\mathfrak{a}$ there exists an integral ideal $\mathfrak{b}$ such that $\mathfrak{a}\mathfrak{b}$ is principal and $\mathfrak{a}\mathfrak{b}$ is nonzero if $\mathfrak{a}$ is.

180β. Show that a nontrivial integral domain is a Dedekind domain if and only if for any two integral ideals $\mathfrak{a}$ and $\mathfrak{b}$, $\mathfrak{a} \subset \mathfrak{b}$ implies $\mathfrak{a} = \mathfrak{b}\mathfrak{c}$ for some integral ideal $\mathfrak{c}$.

180γ. Prove that in a Dedekind domain an ideal is prime if and only if it is not a product of two proper ideals.

181. Proposition. *Every proper prime ideal of a Dedekind domain is a maximal ideal.*

Proof. Suppose $\mathfrak{p}$ is a proper prime ideal of a Dedekind domain R and that $\mathfrak{p} \subset \mathfrak{a}$ where $\mathfrak{a}$ is an integral ideal of R. Then $\mathfrak{a}^{-1}\mathfrak{p} \subset \mathfrak{a}^{-1}\mathfrak{a} = R$, and therefore $\mathfrak{a}^{-1}\mathfrak{p}$ is an integral ideal of R. Now $\mathfrak{a}(\mathfrak{a}^{-1}\mathfrak{p}) = \mathfrak{p}$, and therefore either $\mathfrak{a} \subset \mathfrak{p}$ or $\mathfrak{a}^{-1}\mathfrak{p} \subset \mathfrak{p}$ (**163**). Since $\mathfrak{p} \subset \mathfrak{a}$ by hypothesis, $\mathfrak{a} \subset \mathfrak{p}$ implies $\mathfrak{a} = \mathfrak{p}$. On the other hand, $\mathfrak{a}^{-1}\mathfrak{p} \subset \mathfrak{p}$ implies $\mathfrak{a}^{-1} \subset \mathfrak{p}\mathfrak{p}^{-1} = R$ from which it follows that $R \subset \mathfrak{a}$ and $\mathfrak{a} = R$ (**179α**). We have shown that $\mathfrak{p} \subset \mathfrak{a}$ implies $\mathfrak{a} = \mathfrak{p}$ or $\mathfrak{a} = R$. Therefore $\mathfrak{p}$ is maximal.

Corollary. *If $\mathfrak{a}$ and $\mathfrak{p}$ are proper ideals of a Dedekind domain, $\mathfrak{p}$ is prime, and $\mathfrak{a} \not\subset \mathfrak{p}$, then $\mathfrak{a}\mathfrak{p}^n = \mathfrak{a} \cap \mathfrak{p}^n$ for any $n \in \mathbf{N}$.*

Proof. Since $\mathfrak{a} \not\subset \mathfrak{p}$ and $\mathfrak{p}$ is maximal, $\mathfrak{a} + \mathfrak{p} = (1) = R$. Therefore we have $1 = p + a$ where $p \in \mathfrak{p}$, $a \in \mathfrak{a}$. Consequently

$$1 = 1^n = (p + a)^n = p^n + ra \in \mathfrak{p}^n + \mathfrak{a},$$

where

$$r = \binom{n}{1} p^{n-1} + \binom{n}{2} p^{n-2}a + \cdots + a^{n-1}.$$

If $x \in \mathfrak{a} \cap \mathfrak{p}^n$ then

$$x = x(p^n + ra) = p^n x + rax \in \mathfrak{a}\mathfrak{p}^n,$$

which shows that $\mathfrak{a} \cap \mathfrak{p}^n \subset \mathfrak{a}\mathfrak{p}^n$. Since $\mathfrak{a}\mathfrak{p}^n \subset \mathfrak{a} \cap \mathfrak{p}^n$, equality follows.

181α.   Show that for distinct proper prime ideals $\mathfrak{p}_1, \mathfrak{p}_2, \ldots, \mathfrak{p}_k$ of a Dedekind domain,

$$\mathfrak{p}_1^{\nu_1}\mathfrak{p}_2^{\nu_2} \cdots \mathfrak{p}_k^{\nu_k} = \mathfrak{p}_1^{\nu_1} \cap \mathfrak{p}_2^{\nu_2} \cap \cdots \cap \mathfrak{p}_k^{\nu_k}.$$

182.   Theorem.   *In a Dedekind domain every proper ideal can be factored as a product of proper prime ideals. Furthermore this factorization is unique except for the order of the factors.*

Proof.   We already know that every proper prime ideal of a Dedekind domain is maximal (**181**) and that every nonzero ideal, being invertible, is finitely generated (**179**). (Since (0) is principal, we can say that *every* ideal of a Dedekind domain is finitely generated. Rings with this property are called *Noetherian*.)

Let R be a Dedekind domain. If $\mathfrak{a}$ is a proper ideal of R which is not a product of proper prime ideals of R, then $\mathfrak{a}$ cannot be a prime ideal because we construe "product" to include products of length one. Therefore $\mathfrak{a}$ is not maximal since maximal ideals are prime (**162**). Consequently, there is an ideal $\mathfrak{b}$ of R such that $\mathfrak{a} \subset \mathfrak{b} \subset R$ and $\mathfrak{a} \neq \mathfrak{b} \neq R$. Furthermore, $\mathfrak{a}\mathfrak{b}^{-1} \subset \mathfrak{b}\mathfrak{b}^{-1} = R$ so that $\mathfrak{a}\mathfrak{b}^{-1}$ is an integral ideal. Clearly $\mathfrak{a}$ factors as $(\mathfrak{a}\mathfrak{b}^{-1})\mathfrak{b}$. One of the factors $\mathfrak{a}\mathfrak{b}^{-1}$ or $\mathfrak{b}$ must fail to be a product of proper primes—otherwise $\mathfrak{a}$ would be. Now $\mathfrak{a} \neq \mathfrak{b}$ and $\mathfrak{a} \neq \mathfrak{a}\mathfrak{b}^{-1}$ (why?), and we may say: *if $\mathfrak{a}$ is a proper ideal which is not a product of proper primes, then there exists a strictly larger proper ideal $\mathfrak{a}'$ with the same property.* ($\mathfrak{a}' = \mathfrak{b}$ or $\mathfrak{a}' = \mathfrak{a}\mathfrak{b}^{-1}$.)

By iterated use of this statement, we can derive from the existence of a single ideal with no proper prime factorization the existence of an infinite, ascending chain of ideals of R,

$$\mathfrak{a} \subset \mathfrak{a}' \subset \mathfrak{a}'' \subset \cdots \subset \mathfrak{a}^{(n)} \subset \cdots, \qquad (*)$$

in which $\mathfrak{a}^{(i)} \neq \mathfrak{a}^{(i+1)}$. (See **183**.) We shall see that R does not admit such chains. It is routine to verify that the union

$$\mathfrak{A} = \bigcup_{n=0}^{\infty} \mathfrak{a}^{(n)} = \{r \in R \mid r \in \mathfrak{a}^{(n)}, n \in \mathbf{N}\}$$

is again an ideal of R. Consequently, $\mathfrak{A}$ is finitely generated by elements $a_1, a_2, \ldots, a_k \in R$. For each $i = 1, 2, \ldots, k$ there is a natural number n_i such that $a_i \in \mathfrak{a}^{(n_i)}$. As a result $a_1, a_2, \ldots, a_k \in \mathfrak{a}^{(n)}$ for $n \geq \max\{n_1, n_2, \ldots, n_k\}$. This implies

$$\mathfrak{A} = (a_1, a_2, \ldots, a_k) \subset \mathfrak{a}^{(n)}$$

for large n, and since $\mathfrak{a}^{(n)} \subset \mathfrak{A}$ for *all* n, we have $\mathfrak{a}^{(n)} = \mathfrak{A}$ when n is sufficiently large. Of course this contradicts (for large n) the fact that $\mathfrak{a}^{(n)} \neq \mathfrak{a}^{(n+1)}$. This contradiction shows that we cannot have a proper ideal of R which is not a product of proper primes.

It remains to show uniqueness. We proceed by induction on the number of primes in the factorization, that is, we prove by induction the statements S_n for $n \in \mathbf{N}$.

S_n: If $\mathfrak{a}$ is a proper ideal of R which can be factored as a product of n (or fewer) proper prime ideals, then any two factorizations of $\mathfrak{a}$ are the same except possibly for the order of the factors.

To start the induction, we remark that S_1 is almost obvious. An ideal which is the product of one proper prime ideal is itself a proper prime and has no true factorizations at all. Now we assume S_n and prove S_{n+1}. Suppose that

$$\mathfrak{a} = \mathfrak{p}_1 \mathfrak{p}_2 \cdots \mathfrak{p}_k = \mathfrak{q}_1 \mathfrak{q}_2 \cdots \mathfrak{q}_l$$

are two factorizations of the proper ideal $\mathfrak{a}$ and $k \leq n + 1$. Since $\mathfrak{a} \subset \mathfrak{q}_l$ and $\mathfrak{q}_l$ is prime, one of the factors $\mathfrak{p}_1, \mathfrak{p}_2, \ldots, \mathfrak{p}_k$ must be contained in $\mathfrak{q}_l$ (**163**). We may suppose $\mathfrak{p}_k \subset \mathfrak{q}_l$. Since $\mathfrak{p}_k$ is a proper prime and, consequently, maximal, we must have $\mathfrak{p}_k = \mathfrak{q}_l$. Now, however, we see that

$$\mathfrak{a}\mathfrak{p}_k^{-1} = \mathfrak{p}_1 \mathfrak{p}_2 \cdots \mathfrak{p}_{k-1} = \mathfrak{q}_1 \mathfrak{q}_2 \cdots \mathfrak{q}_{l-1},$$

which falls within the scope of S_n since $k - 1 \leq n$. Therefore these last two factorizations are the same, and the given ones must have been the same.

Corollary. *In a Dedekind domain every proper ideal may be written uniquely in the form* $\mathfrak{p}_1^{v_1} \mathfrak{p}_2^{v_2} \cdots \mathfrak{p}_k^{v_k}$, *where* $\mathfrak{p}_1, \mathfrak{p}_2, \ldots, \mathfrak{p}_k$ *are distinct proper prime ideals and* $v_1, v_2, \ldots, v_k \in \mathbf{N}$.

182α. Prove that in a Dedekind domain every proper *fractionary* ideal can be written uniquely in the form $\mathfrak{p}_2^{v_1} \mathfrak{p}_2^{v_2} \cdots \mathfrak{p}_k^{v_k}$, where $\mathfrak{p}_1, \mathfrak{p}_2, \ldots, \mathfrak{p}_k$ are distinct prime ideals and $v_1, v_2, \ldots, v_k$ are *nonzero* integers.

182β. Prove that in a Dedekind domain every proper (integral) ideal can be written uniquely in the form $\mathfrak{p}_1^{v_1} \cap \mathfrak{p}_2^{v_2} \cap \cdots \cap \mathfrak{p}_k^{v_k}$, where $\mathfrak{p}_1, \mathfrak{p}_2, \ldots, \mathfrak{p}_k$ are distinct prime ideals and $v_1, v_2, \ldots, v_k \in \mathbf{N}$.

182γ. Let $\mathfrak{a}$ and $\mathfrak{b}$ be fractionary ideals of a Dedekind domain and $\mathfrak{p}_1, \mathfrak{p}_2, \ldots,$ $\mathfrak{p}_k$ be prime ideals. Suppose further that

$$\mathfrak{a} = \mathfrak{p}_1^{v_1} \mathfrak{p}_2^{v_2} \cdots \mathfrak{p}_k^{v_k} \quad \text{and} \quad \mathfrak{b} = \mathfrak{p}_1^{\mu_1} \mathfrak{p}_2^{\mu_2} \cdots \mathfrak{p}_k^{\mu_k}.$$

Show that $\mathfrak{a} \subset \mathfrak{b}$ if and only if $\nu_i \geq \mu_i$ for $i = 1, 2, \ldots, k$. (Here the ν_i's and μ_i's are allowed to be any integers.)

182δ. Using the notation of the previous problem, show that

$$\mathfrak{a} \cap \mathfrak{b} = \mathfrak{p}_1^{\max(\nu_1, \mu_1)} \mathfrak{p}_2^{\max(\nu_2, \mu_2)} \cdots \mathfrak{p}_k^{\max(\nu_k, \mu_k)},$$

$$\mathfrak{a}\mathfrak{b} = \mathfrak{p}_1^{\nu_1 + \mu_1} \mathfrak{p}_2^{\nu_2 + \mu_2} \cdots \mathfrak{p}_k^{\nu_k + \mu_k},$$

$$\mathfrak{a} + \mathfrak{b} = \mathfrak{p}_1^{\min(\nu_1, \mu_1)} \mathfrak{p}_2^{\min(\nu_2, \mu_2)} \cdots \mathfrak{p}_k^{\min(\nu_k, \mu_k)}.$$

182ε. Show that for any three fractionary ideals $\mathfrak{a}$, $\mathfrak{b}$, $\mathfrak{c}$ of a Dedekind domain,

$$\mathfrak{a} + (\mathfrak{b} \cap \mathfrak{c}) = (\mathfrak{a} + \mathfrak{b}) \cap (\mathfrak{a} + \mathfrak{c}),$$

$$\mathfrak{a} \cap (\mathfrak{b} + \mathfrak{c}) = (\mathfrak{a} \cap \mathfrak{b}) + (\mathfrak{a} \cap \mathfrak{c}).$$

183. *The Axiom of Choice.* The preceding proof depends upon a set-theoretic principle, called the axiom of choice, which has a different character and a higher order than the simple rules of intuitive set theory which have served us to this point. It is worth a digression to make explicit this principle and its use in the present instance, the only place we need it.

Axiom of Choice.   For any set X there exists a mapping

$$\phi: 2^X \to X$$

such that $\phi A \in A$ for each nonempty subset A of X. (Recall that 2^X denotes the power set of X defined in **14**.) The mapping ϕ is called a *choice function* for X because it "chooses" an element ϕA from each nonempty subset A.

In the proof of **182** we must apply the axiom of choice to the set $\mathscr{I}_R$ of ideals of R in order to construct the ascending chain (∗). Let $\phi: 2^{\mathscr{I}_R} \to \mathscr{I}_R$ be a choice function for $\mathscr{I}_R$. If $\mathfrak{a}$ is any ideal without a proper prime factorization, let $\mathscr{C}(\mathfrak{a})$ denote the class of ideals which contain $\mathfrak{a}$ properly and which themselves have no proper prime factorization. The first argument in **182** shows that for such $\mathfrak{a}$, $\mathscr{C}(\mathfrak{a})$ is not empty. Therefore the chain (∗) is defined inductively by the rule $\mathfrak{a}^{(n+1)} = \phi(\mathscr{C}(\mathfrak{a}^{(n)}))$.

184. *The Chinese Remainder Theorem.* Let $\mathfrak{a}_1$, $\mathfrak{a}_2$, $\ldots$, $\mathfrak{a}_n$ *be ideals of a Dedekind domain R and $a_1, a_2, \ldots, a_n$ elements of R. There exists an element $x \in R$, such that*

$$x \equiv a_i \bmod \mathfrak{a}_i, \qquad i = 1, 2, \ldots, n, \tag{∗}$$

if and only if $a_i \equiv a_j \bmod (\mathfrak{a}_i + \mathfrak{a}_j)$ for all i and j.

Proof. The conditions $a_i \equiv a_j \bmod (\mathfrak{a}_i + \mathfrak{a}_j)$ are clearly necessary. The proof of their sufficiency proceeds by induction on n beginning with $n = 2$.

For $n = 2$, $a_1 \equiv a_2 \bmod (\mathfrak{a}_1 + \mathfrak{a}_2)$ implies that

$$a_1 - a_2 = a_1' + a_2' \in \mathfrak{a}_1 + \mathfrak{a}_2,$$

where $a_1' \in \mathfrak{a}_1$, $a_2' \in \mathfrak{a}_2$. Then $x = a_1 - a_1' = a_2 + a_2'$ has the required properties.

For $n > 2$, we may suppose, as a result of the induction hypothesis, that there is an element $x' \in R$ such that $x' \equiv a_i \bmod \mathfrak{a}_i$ for $i = 1, 2, \ldots, n - 1$. Let

$$\mathfrak{a} = \mathfrak{a}_1 \cap \mathfrak{a}_2 \cap \cdots \cap \mathfrak{a}_{n-1}.$$

Then x will be a complete solution to the set of n congruences (∗) if $x \equiv x'$ mod $\mathfrak{a}$ and $x \equiv a_n \bmod \mathfrak{a}_n$. There exists such an $x \in R$ provided that $x' \equiv x_n$ mod $(\mathfrak{a} + \mathfrak{a}_n)$. However,

$$\mathfrak{a} + \mathfrak{a}_n = \left(\bigcap_{i=1}^{n-1} \mathfrak{a}_i \right) + \mathfrak{a}_n = \bigcap_{i=1}^{n-1} (\mathfrak{a}_i + \mathfrak{a}_n),$$

which follows from **182ε**. Since $x' \equiv x_i \bmod \mathfrak{a}_i$ for $i = 1, 2, \ldots, n - 1$, we have

$$x' \equiv x_i \equiv x_n \bmod (\mathfrak{a}_i + \mathfrak{a}_n) \quad \text{for } i = 1, 2, \ldots, n - 1.$$

This means that for $i = 1, 2, \ldots, n - 1$,

$$x' - x_n \in \mathfrak{a}_i + \mathfrak{a}_n$$

and therefore

$$x' - x_n \in \mathfrak{a} + \mathfrak{a}_n.$$

Thus the condition $x' \equiv x_n$ is satisfied and (∗) has a solution $x \in R$.

As an example, this theorem implies that the set of congruences

$$x \equiv 1 \bmod 6, \qquad x \equiv 5 \bmod 8, \qquad x \equiv 4 \bmod 9$$

has a solution. Indeed any integer $x \equiv 13 \bmod 72$ is a solution. The name of this theorem recalls its use (for integers) by the Chinese astronomers of ancient times in calender reckoning.

184α. Let R be a Dedekind domain and $\mathfrak{a}$ a proper ideal of R with factorization $\mathfrak{a} = \mathfrak{p}_1^{v_1} \mathfrak{p}_2^{v_2} \cdots \mathfrak{p}_k^{v_k}$. Prove that $R/\mathfrak{a}$ is isomorphic to the direct product $(R/\mathfrak{p}_1^{v_1}) \times (R/\mathfrak{p}_2^{v_2}) \times \cdots \times (R/\mathfrak{p}_k^{v_k})$.

184β.   Let $a_1, a_2, \ldots, a_n$ be distinct elements of a field F. Using the Chinese remainder theorem, show that there exists a polynomial f over F which takes given values $f(a_i) = c_i$ for $i = 1, 2, \ldots, n$. (Compare **100α**.)

184γ.   Prove that a Dedekind domain with a finite number of prime ideals is a principal ideal domain.

184δ.   Prove that every ideal of a Dedekind domain can be generated by two of its elements, one of which may be chosen arbitrarily.

184ε.   Let $\mathfrak{a}$ and $\mathfrak{b}$ be ideals of a Dedekind domain. Show that there exists a principal ideal (c) such that $(c) + \mathfrak{a}\mathfrak{b} = \mathfrak{a}$.

185.   In the next articles we shall develop a proof of the converse of the unique factorization theorem for ideals in a Dedekind domain (**182**). It will be convenient to call an ideal *primigenial* if it is a product of proper prime ideals. A *primigenial ring* is one in which every proper ideal is primigenial.

Proposition.   *In a nontrivial integral domain factorization of invertible primigenial ideals is unique.*

Proof.   Suppose $\mathfrak{a}$ is an invertible ideal of a nontrivial integral domain, and

$$\mathfrak{a} = \mathfrak{p}_1 \mathfrak{p}_2 \cdots \mathfrak{p}_k = \mathfrak{q}_1 \mathfrak{q}_2 \cdots \mathfrak{q}_l$$

are two factorizations of $\mathfrak{a}$ as a product of proper primes. Each of the $\mathfrak{p}_i$'s and $\mathfrak{q}_j$'s is invertible (**179γ**); in fact we have

$$\mathfrak{p}_i^{-1} = \mathfrak{a}^{-1}\mathfrak{p}_1 \cdots \mathfrak{p}_{i-1}\mathfrak{p}_{i+1} \cdots \mathfrak{p}_k$$

and a similar formula for $\mathfrak{q}_j^{-1}$. Among the primes $\mathfrak{p}_1, \mathfrak{p}_2, \ldots, \mathfrak{p}_k$ choose one which is minimal in the sense that it does not contain any of the others properly. We may assume that this is $\mathfrak{p}_k$. Since $\mathfrak{a} = \mathfrak{q}_1 \mathfrak{q}_2 \cdots \mathfrak{q}_l$ is contained in $\mathfrak{p}_k$, a prime, at least one $\mathfrak{q}_i$ is contained in $\mathfrak{p}_k$ (**163**). We may suppose $\mathfrak{q}_l \subset \mathfrak{p}_k$. Similarly, $\mathfrak{p}_1 \mathfrak{p}_2 \cdots \mathfrak{p}_k \subset \mathfrak{q}_l$ implies $\mathfrak{p}_i \subset \mathfrak{q}_l$ for some i. However, now we have $\mathfrak{p}_i \subset \mathfrak{q}_l \subset \mathfrak{p}_k$, and hence $\mathfrak{p}_i \subset \mathfrak{p}_k$. The choice of $\mathfrak{p}_k$ forces $\mathfrak{p}_i = \mathfrak{p}_k$, and then $\mathfrak{p}_k \subset \mathfrak{q}_l \subset \mathfrak{p}_k$ implies $\mathfrak{q}_l = \mathfrak{p}_k$. Now we have

$$\mathfrak{a}' = \mathfrak{a}\mathfrak{p}_k^{-1} = \mathfrak{a}\mathfrak{q}_l^{-1} = \mathfrak{p}_1 \mathfrak{p}_2 \cdots \mathfrak{p}_{k-1} = \mathfrak{q}_1 \mathfrak{q}_2 \cdots \mathfrak{q}_{l-1}.$$

The ideal $\mathfrak{a}'$ is invertible because it is a product of invertible ideals. Therefore the entire argument may be repeated with $\mathfrak{a}'$ in place of $\mathfrak{a}$. The conclusion follows in a finite number of steps of this kind.

186.   Proposition.   *If R is a primigenial ring and $\mathfrak{p}$ is a proper prime ideal of R, then the quotient ring $\bar{R} = R/\mathfrak{p}$ is a primigenial ring.*

Proof. Since $\mathfrak{p}$ is a proper prime ideal, $\bar{R}$ is a nontrivial integral domain. Let $\phi : R \to \bar{R}$ denote the canonical epimorphism, given by $\phi a = a + \mathfrak{p} = \bar{a}$. Since ϕ is an epimorphism, it carries ideals of R to ideals of $\bar{R}$ and preserves products of ideals, that is, $\phi(\mathfrak{ab}) = (\phi\mathfrak{a})(\phi\mathfrak{b})$. (We leave the proof to the reader.)

Furthermore, ϕ gives a one-to-one correspondence between the proper prime ideals of $\bar{R}$ and the proper prime ideals of R which contain $\mathfrak{p}$. Indeed, if $\bar{\mathfrak{q}}$ is a proper prime ideal of $\bar{R}$, then $\phi^{-1}(\bar{\mathfrak{q}})$ is a proper prime ideal of R containing $\mathfrak{p}$, as we have previously shown (**166**). On the other hand, if $\mathfrak{q}$ is a proper prime ideal of R and $\bar{a}\bar{b} \in \phi(\mathfrak{q})$, then $ab \in \mathfrak{q}$, where $\phi a = \bar{a}, \phi b = \bar{b}$. Consequently, $a \in \mathfrak{q}$ or $b \in \mathfrak{q}$, from which is follows that $\bar{a} \in \phi(\mathfrak{q})$ or $\bar{b} \in \phi(\mathfrak{q})$, and thus $\phi(\mathfrak{q})$ is prime.

Now suppose that $\bar{a}$ is a proper ideal of $\bar{R}$. Then $\phi^{-1}(\bar{a})$ is a proper ideal of R and has a factorization $\phi^{-1}(\bar{a}) = \mathfrak{p}_1\mathfrak{p}_2 \cdots \mathfrak{p}_k$ as a product of proper primes each of which contains $\mathfrak{p}$. Therefore in $\bar{R}$ we have the factorization

$$\bar{a} = \phi\phi^{-1}(\bar{a}) = (\phi\mathfrak{p}_1)(\phi\mathfrak{p}_2) \cdots (\phi\mathfrak{p}_k).$$

187. Theorem. *A primigenial ring is a Dedekind domain.*

Proof. Let R be a primigenial ring. The essential point is to show that *every invertible proper prime ideal $\mathfrak{p}$ of R is maximal.* Suppose that $\mathfrak{p}$ is an invertible prime ideal and $(0) \neq \mathfrak{p} \neq (1)$. Let $a \in R - \mathfrak{p}$. Then the ideals $\mathfrak{p} + (a)$ and $\mathfrak{p}^2 + (a)$ have prime factorizations

$$\mathfrak{p} + (a) = \mathfrak{p}_1\mathfrak{p}_2 \cdots \mathfrak{p}_k \quad \text{and} \quad \mathfrak{p}^2 + (a) = \mathfrak{q}_1\mathfrak{q}_2 \cdots \mathfrak{q}_l .$$

Clearly each $\mathfrak{p}_i$ contains $\mathfrak{p}$, but further, we have

$$\mathfrak{p}\mathfrak{p}_1\mathfrak{p}_2 \cdots \mathfrak{p}_k = \mathfrak{p}(\mathfrak{p} + (a)) = \mathfrak{p}^2 + \mathfrak{p}(a) \subset \mathfrak{p}^2 + (a) \subset \mathfrak{q}_j ,$$

which implies that each $\mathfrak{q}_j$ contains one of the factors of $\mathfrak{p}^2 + \mathfrak{p}(a)$. Each of these factors, $\mathfrak{p}, \mathfrak{p}_1, \mathfrak{p}_2, \ldots, \mathfrak{p}_k$, contains $\mathfrak{p}$, and therefore each prime $\mathfrak{q}_j$ contains $\mathfrak{p}$. Passing to the quotient ring $\bar{R} = R/\mathfrak{p}$, which is also primigenial by **186**, we see that

$$\bar{\mathfrak{p}}_1\bar{\mathfrak{p}}_2 \cdots \bar{\mathfrak{p}}_k = \bar{\mathfrak{p}} + (\bar{a}) = (\bar{a}) = \bar{\mathfrak{p}}^2 + (\bar{a}) = \bar{\mathfrak{q}}_1\bar{\mathfrak{q}}_2 \cdots \bar{\mathfrak{q}}_l , \qquad (*)$$

where bars denote images under the canonical epimorphism $\phi : R \to \bar{R}$. Since $(\bar{a})$ is principal and therefore invertible, the two factorizations of $(\bar{a})$ in $(*)$ must be identical except for order (**185**), and renumbering if necessary, we may assume that $\bar{\mathfrak{p}}_i = \bar{\mathfrak{q}}_i$ for $i = 1, 2, \ldots, l = k$. Since all the $\mathfrak{p}_i$'s and $\mathfrak{q}_i$'s contain $\mathfrak{p}$, it follows from $\bar{\mathfrak{p}}_i = \bar{\mathfrak{q}}_i$ that $\mathfrak{p}_i = \mathfrak{q}_i$ for each i. As a result we have $\mathfrak{p} + (a) = \mathfrak{p}^2 + (a)$, from which we conclude that $\mathfrak{p} \subset \mathfrak{p}^2 + (a)$. Consequently,

$$\mathfrak{p} = \mathfrak{p} \cap (\mathfrak{p}^2 + (a)) \subset \mathfrak{p}^2 + (a)\mathfrak{p} \subset \mathfrak{p}.$$

Therefore $p = p^2 + (a)p$, and since p is invertible, this implies

$$R = p^{-1}p = p^{-1}(p^2 + (a)p) = p + (a).$$

Because a was any element of $R - p$, this shows that p is maximal. Thus we have established the essential point.

The rest of the proof is easy. Let p be any proper prime ideal of R and $a \in p$. Then $(a) = p_1 p_2 \cdots p_k \subset p$ for primes p_i, each of which is invertible and consequently maximal. However, one of the p_i's must be contained in p since their product is. Now $p_i \subset p$ implies $p = p_i$ (since p_i is maximal), and we have shown that any prime ideal is invertible. Since every proper ideal of R is a product of prime ideals, it follows that all such ideals are invertible and R is a Dedekind domain.

Integral Extensions

---

188.   Let R be a subring of a ring R'. An element $x \in R'$ is said to be *integral over* R if it satisfies an equation of the form

$$x^n + a_1 x^{n-1} + \cdots + a_{n-1}x + a_n = 0$$

in which the leading coefficient is 1 and the other coefficients, $a_1, a_2, \ldots, a_n,$ are elements of R. The set of elements of R' which are integral over R is called the *integral closure of R in R'*, and will be denoted $\bar{R}$. When $\bar{R} = R$, we say that R is *integrally closed in R'*. When $\bar{R} = R'$, we say that R' is an *integral extension of R*.

189.   A ring R' is a *finite extension* of a subring R if there exist elements $z_1, z_2, \ldots, z_n \in R'$ such that every element $z \in R'$ can be written in the form

$$z = r_1 z_1 + r_2 z_2 + \cdots + r_n z_n$$

for some elements $r_1, r_2, \ldots, r_n \in R$. In this case the set $\{z_1, z_2, \ldots, z_n\}$ is called a *basis* for R' over R.

Proposition.   *If R' is a nontrivial integral domain which is a finite extension of a subring R, then R' is an integral extension of R.*

Proof. Let $\{z_1, z_2, \ldots, z_n\}$ be a basis for R' over R. Then for any $z \in R'$ we have $zz_i = \sum_{j=1}^{n} r_{ij} z_j$, where $r_{ij} \in R$ for $i, j = 1, 2, \ldots, n$. Thus we may view $z_1, z_2, \ldots, z_n$ as the solution of the system of linear equations over $\mathbf{Q}_{R'}$ in unknowns $X_1, X_2, \ldots, X_n$:

$$\begin{cases} (r_{11} - z)X_1 + r_{12} X_2 + \cdots + r_{1n} X_n = 0 \\ r_{21} X_1 + (r_{22} - z)X_2 + \cdots + r_{2n} X_n = 0 \\ \ \vdots \qquad\qquad \vdots \qquad\qquad\quad \vdots \\ r_{n1} X_1 + r_{n2} X_2 + \cdots + (r_{nn} - z)X_n = 0. \end{cases} \tag{*}$$

Therefore the determinant

$$\begin{vmatrix} r_{11} - z & r_{12} & \cdots & r_{1n} \\ r_{21} & r_{22} - z & \cdots & r_{2n} \\ \vdots & \vdots & & \vdots \\ r_{n1} & r_{n2} & \cdots & r_{nn} - z \end{vmatrix} = 0.$$

However, this determinant is just a polynomial in z with coefficients in R and with leading coefficient $(-1)^n$, which shows that z is integral over R. Thus every element $z \in R'$ is integral over R, and R' is an integral extension of R.

189α. Show that if R' is a finite extension of R and R'' is a finite extension of R', that R'' is a finite extension of R.

190. If R is a subring of R' and $x \in R'$, we denote by $R[x]$ the smallest subring of R' containing both R and x. Clearly $R[x]$ consists of all the elements of R' which can be written as polynomials in x with coefficients in R.

Proposition. *An element x of a nontrivial integral domain R' is integral over a subring R if and only if $R[x]$ is a finite extension of R.*

Proof. By the preceding proposition if $R[x]$ is a finite extension of R, then it is an integral extension of R, and consequently $x \in R[x]$ is integral over R. On the other hand, if x is integral over R and satisfies the equation

$$x^n + a_1 x^{n-1} + \cdots + a_{n-1} x + a_n = 0, \qquad a_1, a_2, \ldots, a_n \in R,$$

then every element of $R[x]$ may be written as a polynomial in x of degree less than n. (How?) Therefore $\{1, x, x^2, \ldots, x^{n-1}\}$ is a basis for $R[x]$ over R.

191. *Proposition.* *The integral closure $\bar{R}$ of a subring R of a nontrivial integral domain R' is a subring of R'. Furthermore $\bar{R}$ is integrally closed in R'.*

Proof. Suppose $x, y \in \overline{R}$. Then $R[x]$ is a finite extension of R, and $R[x, y] = (R[x])[y]$ is a finite extension of $R[x]$ because x is integral over R and y is integral over $R[x]$. It follows that $R[x, y]$ is a finite extension of R. (See **189α.**) In fact, if x and y satisfy equations of degree n and m respectively over R, then the set of nm elements $x^i y^j$ where $0 \le i < n$ and $0 \le j < m$ is a basis for $R[x, y]$ over R. Since $R[x, y]$ is a finite extension of R, it is an integral extension (**189**); consequently, the elements $x - y$ and xy of $R[x, y]$ are integral over R. This shows that $\overline{R}$ is a ring.

Suppose now that x is an element of R' and that x is integral over $\overline{R}$. Then x satisfies an equation

$$x^n + \bar{a}_1 x^{n-1} + \cdots + \bar{a}_{n-1} x + \bar{a}_n = 0, \qquad \bar{a}_1, \bar{a}_2, \ldots, \bar{a}_n \in \overline{R}.$$

It follows that we have a sequence of finite extensions

$$R \subset R[\bar{a}_1] \subset R[\bar{a}_1, \bar{a}_2] \subset \cdots \subset R[\bar{a}_1, \bar{a}_2, \ldots, \bar{a}_n] \subset R[\bar{a}_1, \bar{a}_2, \ldots, \bar{a}_n, x].$$

Therefore $R[\bar{a}_1, \bar{a}_2, \ldots, \bar{a}_n, x]$ is a finite extension, and consequently, an integral extension of R. Thus x is integral over R, that is, $x \in \overline{R}$.

Algebraic Integers

192. We recall that an *algebraic number* is a complex number which is algebraic over the rational field $\mathbf{Q}$ and that the set of all such numbers forms a field $\overline{\mathbf{Q}}$ (**108**). Analogously, an *algebraic integer* is a complex number which is integral over the ring of integers $\mathbf{Z}$. By the preceding proposition we know that the set of algebraic integers forms a ring $\overline{\mathbf{Z}}$. Of course $\overline{\mathbf{Z}}$ is just the integral closure of $\mathbf{Z}$ considered as a subring of $\mathbf{C}$. Furthermore, since every element of $\overline{\mathbf{Z}}$ satisfies an algebraic equation over $\mathbf{Z}$, it is clear that an algebraic integer is an algebraic number, that is, $\overline{\mathbf{Z}} \subset \overline{\mathbf{Q}}$.

In general the algebraic integers which belong to a given number field F are called the *integers of F* and form a ring which we denote $\mathbf{Z}_F$. Obviously $\mathbf{Z}_F = \overline{\mathbf{Z}} \cap F$. By definition we have $\mathbf{Z}_\mathbf{C} = \overline{\mathbf{Z}}$.

Proposition. *The ring of integers $\mathbf{Z}$ is integrally closed in the rational field $\mathbf{Q}$, that is to say, $\mathbf{Z}_\mathbf{Q} = \mathbf{Z}$.*

Proof. Suppose that $r/s \in \mathbf{Q}$ is an algebraic integer where $r, s \in \mathbf{Z}$ and $(r, s) = 1$. Then for some $a_1, a_2, \ldots, a_n \in \mathbf{Z}$,

$$\left(\frac{r}{s}\right)^n + a_1 \left(\frac{r}{s}\right)^{n-1} + \cdots + a_{n-1} \left(\frac{r}{s}\right) + a_n = 0 \tag{1}$$

and

$$r^n + a_1 r^{n-1} s + \cdots + a_{n-1} r s^{n-1} + a_n s^n = 0. \tag{2}$$

Equation (2) shows that $s \mid r^n$. Since $(r, s) = 1$, we must have $s = \pm 1$ and $r/s = \pm r \in \mathbf{Z}$. This completes the proof.

The elements of $\mathbf{Z}$ are often called *rational integers* to emphasize that they are the integers of the rational field $\mathbf{Q}$.

Proposition. *For every algebraic number α there exists a rational integer m such that $m\alpha$ is an algebraic integer.*

Proof. We may assume without loss of generality that α is a root of a primitive polynomial f given by

$$fx = a_0 x^n + a_1 x^{n-1} + \cdots + a_{n-1} x + a_n, \qquad a_0, a_1, \ldots, a_n \in \mathbf{Z}.$$

Then $a_0 \alpha$ is a root of the polynomial g given by

$$gx = x^n + a_1 x^{n-1} + \cdots + a_{n-1} a_0^{n-2} x + a_n a_0^{n-1}.$$

Since g has integral coefficients, $a_0 \alpha$ is an algebraic integer, and we may take $m = a_0$.

192α. Let R be a number domain (subring of $\mathbf{C}$) which is a principal ideal domain. Show that R is integrally closed in its field of fractions $\mathbf{Q}_R$.

192β. Show that the ring of integers of the field $\mathbf{Q}(i)$ is $\mathbf{Z}(i)$, the ring of Gaussian integers.

192γ. Show that the polynomial ring $\mathbf{Z}[x]$ is integrally closed in its field of fractions.

192δ. Let α be an algebraic integer of E, a Galois extension of $\mathbf{Q}$ with group $\mathscr{G}(E/\mathbf{Q}) = \{\phi_1, \phi_2, \ldots, \phi_n\}$. Prove that all the conjugates $\phi_1 \alpha, \phi_2 \alpha, \ldots, \phi_n \alpha$ of α are algebraic integers and that the polynomial

$$(x - \phi_1 \alpha)(x - \phi_2 \alpha) \cdots (x - \phi_n \alpha)$$

has coefficients which are rational integers.

193. Proposition. *If f is a polynomial over $\overline{\mathbf{Z}}$ and α is a root of f, then $(fx)/(x - \alpha)$ is also a polynomial over $\overline{\mathbf{Z}}$.*

Proof.   The proof is by induction on the degree of f. When $\deg f = 1$, we have $fx = a(x - \alpha)$ where $a \in \mathbf{Z}$ and the result is obvious. To accomplish the induction step, we suppose the statement holds for polynomials of degree less than n. Suppose $\deg f = n$ and

$$fx = a_0 x^n + a_1 x^{n-1} + \cdots + a_{n-1} x + a_n.$$

It follows from the argument of the preceding article that:

(1) $a_0 \in \overline{\mathbf{Z}}$,
(2) the polynomial g defined by $fx = a_0 x^{n-1}(x - \alpha) + gx$ has coefficients in $\overline{\mathbf{Z}}$,
(3) $\deg g < n$. (Why?)

Therefore, by the induction hypothesis $(gx)/(x - \alpha)$ has coefficients in $\overline{\mathbf{Z}}$, and consequently so does f since

$$\left(\frac{fx}{x - \alpha}\right) = a_0 x^{n-1} + \left(\frac{gx}{x - \alpha}\right).$$

Corollary.   *If $\alpha_1, \alpha_2, \ldots, \alpha_k$ are roots of the polynomial f over $\overline{\mathbf{Z}}$ given by*

$$fx = a_0 x^n + a_1 x^{n-1} + \cdots + a_{n-1} x + a_n,$$

then $a_0 \alpha_1 \alpha_2 \cdots \alpha_k$ is an algebraic integer.

Proof.   Let $\alpha_{k+1}, \alpha_{k+2}, \ldots, \alpha_n$ denote the remaining roots of f. Applying the proposition $n - k$ times shows that the polynomial

$$\frac{f(x)}{(x - \alpha_{k+1})(x - \alpha_{k+2}) \cdots (x - \alpha_n)} = a_0(x - \alpha_1)(x - \alpha_2) \cdots (x - \alpha_n)$$

has coefficients in $\overline{\mathbf{Z}}$. However, $a_0 \alpha_1 \alpha_2 \cdots \alpha_k$ is the constant term of this polynomial.

194.   In this and the following two articles we shall consider the situation where R is the ring of algebraic integers of a number field E which is a Galois extension of the rational field $\mathbf{Q}$. To fix notation, we assume that $[E : \mathbf{Q}] = n$ and that the Galois group is $\mathscr{G}(E/\mathbf{Q}) = \{\phi_1, \phi_2, \ldots, \phi_n\}$. We shall omit repetition of these assumptions in the hypotheses of the propositions.

Proposition.   *There exists a basis $\omega_1, \omega_2, \ldots, \omega_n$ for E over $\mathbf{Q}$ such that every element $\alpha \in R$ can be written uniquely in the form*

$$\alpha = a_1 \omega_1 + a_2 \omega_2 + \cdots + a_n \omega_n,$$

where $a_1, a_2, \ldots, a_n \in \mathbf{Z}$.

Proof. Since every algebraic number can be multiplied by an integer to obtain an algebraic integer, we may begin with the assumption that we have a basis $\bar{\omega}_1, \bar{\omega}_2, \ldots, \bar{\omega}_n$ for E over $\mathbf{Q}$ consisting of algebraic integers. Now we form the determinant

$$\delta = \begin{vmatrix} \phi_1\bar{\omega}_1 & \phi_1\bar{\omega}_2 & \cdots & \phi_1\bar{\omega}_n \\ \phi_2\bar{\omega}_1 & \phi_2\bar{\omega}_2 & \cdots & \phi_2\bar{\omega}_n \\ \vdots & \vdots & & \vdots \\ \phi_n\bar{\omega}_1 & \phi_n\bar{\omega}_2 & \cdots & \phi_n\bar{\omega}_n \end{vmatrix}.$$

We observe that δ is an algebraic integer, or more specifically, that $\delta \in R$ since each of the elements $\phi_i\bar{\omega}_j$ belongs to R. (**192δ**). Furthermore, for $\phi \in \mathcal{G}(E/\mathbf{Q})$ we have

$$\phi\delta = \begin{vmatrix} \phi\phi_1\bar{\omega}_1 & \phi\phi_1\bar{\omega}_2 & \cdots & \phi\phi_1\bar{\omega}_n \\ \phi\phi_2\bar{\omega}_1 & \phi\phi_2\bar{\omega}_2 & \cdots & \phi\phi_2\bar{\omega}_n \\ \vdots & \vdots & & \vdots \\ \phi\phi_n\bar{\omega}_1 & \phi\phi_n\bar{\omega}_2 & \cdots & \phi\phi_n\bar{\omega}_n \end{vmatrix}.$$

However, the effect of letting ϕ act this way is simply to permute the rows. Consequently, $\phi\delta = \pm\delta$ for any $\phi \in \mathcal{G}(E/\mathbf{Q})$. Therefore $\delta^2 \in \mathbf{Q}$. Since δ^2 is an algebraic integer, it follows that $\delta^2 \in \mathbf{Z}$. Now we set $\omega_i = \bar{\omega}_i/\delta^2$. Clearly $\omega_1, \omega_2, \ldots, \omega_n$ form a basis for E over $\mathbf{Q}$.

Suppose now that $\alpha \in R$ and that we have written α as

$$\alpha = \bar{a}_1\bar{\omega}_1 + \bar{a}_2\bar{\omega}_2 + \cdots + \bar{a}_n\bar{\omega}_n,$$

where $\bar{a}_1, \bar{a}_2, \ldots, \bar{a}_n \in \mathbf{Q}$. Applying $\phi_1, \phi_2, \ldots, \phi_n$ to α, we obtain:

$$\begin{cases} \phi_1\alpha = \bar{a}_1(\phi_1\bar{\omega}_1) + \bar{a}_2(\phi_1\bar{\omega}_2) + \cdots + \bar{a}_n(\phi_1\bar{\omega}_n) \\ \phi_2\alpha = \bar{a}_1(\phi_2\bar{\omega}_1) + \bar{a}_2(\phi_2\bar{\omega}_2) + \cdots + \bar{a}_n(\phi_2\bar{\omega}_n) \\ \vdots \qquad \vdots \qquad \vdots \qquad \vdots \\ \phi_n\alpha = \bar{a}_1(\phi_n\bar{\omega}_1) + \bar{a}_2(\phi_n\bar{\omega}_2) + \cdots + \bar{a}_n(\phi_n\bar{\omega}_n). \end{cases} \qquad (*)$$

We may therefore interpret $\bar{a}_1, \bar{a}_2, \ldots, \bar{a}_n$ as the solution of the system $(*)$ of linear equations over E. Consequently $\bar{a}_i = \delta_i/\delta$, where δ is the determinant above and where

$$\delta_i = \begin{matrix} (1) \quad\;\; (2) \quad\; \cdots \quad\; (i) \quad\; \cdots \quad\; (n) \\ \begin{vmatrix} \phi_1\bar{\omega}_1 & \phi_1\bar{\omega}_2 & \cdots & \phi_1\alpha & \cdots & \phi_1\bar{\omega}_n \\ \phi_2\bar{\omega}_1 & \phi_2\bar{\omega}_2 & \cdots & \phi_2\alpha & \cdots & \phi_2\bar{\omega}_n \\ \vdots & \vdots & & \vdots & & \vdots \\ \phi_n\bar{\omega}_1 & \phi_n\bar{\omega}_2 & \cdots & \phi_n\alpha & \cdots & \phi_n\bar{\omega}_n \end{vmatrix} \end{matrix}$$

is obtained from δ by modifying the i-th column. Clearly δ_i is an algebraic integer. Furthermore $\delta^2\bar{a}_i = \delta\delta_i \in R \cap \mathbf{Q} = \mathbf{Z}$. Now we set $a_i = \delta^2\bar{a}_i \in \mathbf{Z}$, and we have

$$\alpha = \bar{a}_1\bar{\omega}_1 + \bar{a}_2\bar{\omega}_2 + \cdots + \bar{a}_n\bar{\omega}_n$$

$$= \delta^2\bar{a}_1\frac{\bar{\omega}_1}{\delta^2} + \delta^2\bar{a}_2\frac{\bar{\omega}_2}{\delta^2} + \cdots + \delta^2\bar{a}_n\frac{\bar{\omega}_n}{\delta^2}$$

$$= a_1\omega_1 + a_2\omega_2 + \cdots + a_n\omega_n.$$

Since $\omega_1, \omega_2, \ldots, \omega_n$ form a basis for E over $\mathbf{Q}$, this final expression for α must be unique.

194α. Prove the proposition above assuming only that E is a finite extension of $\mathbf{Q}$.

195. Proposition. *If $\mathfrak{a}$ is an ideal of R, then there exist elements $\alpha_1, \alpha_2, \ldots,$ $\alpha_n \in \mathfrak{a}$ such that every element $\alpha \in \mathfrak{a}$ may be written in the form*

$$\alpha = c_1\alpha_1 + c_2\alpha_2 + \cdots + c_n\alpha_n,$$

where $c_1, c_2, \ldots, c_n \in \mathbf{Z}$.

Proof. We define mappings $f_1, f_2, \ldots, f_n$ from $\mathfrak{a}$ to $\mathbf{Z}$ by $f_i\alpha = a_i$, where $\alpha = a_1\omega_1 + a_2\omega_2 + \cdots + a_n\omega_n$ is the unique expression of α in terms of a basis $\omega_1, \omega_2, \ldots, \omega_n$, chosen as guaranteed by the preceding article. The f_i are not ring homomorphisms in general, but they are homomorphisms of the additive group structure. Therefore, the sets

$$A_1 = \text{Im} f_1 = f_1(\mathfrak{a}),$$
$$A_2 = f_2(\text{Ker} f_1),$$
$$\vdots$$
$$A_n = f_n(\text{Ker} f_1 \cap \text{Ker} f_2 \cap \cdots \cap \text{Ker} f_{n-1})$$

are subgroups of $\mathbf{Z}$. Therefore there are integers $k_1, k_2, \ldots, k_n$ such that $A_i = k_i\mathbf{Z}$. It follows that we can choose elements $\alpha_1, \alpha_2, \ldots, \alpha_n \in \mathfrak{a}$ such that $f_i\alpha_i = k_i$ and $f_j\alpha_i = 0$ for $j < i$.

Given $\alpha \in \mathfrak{a}$, we have $f_1\alpha = c_1k_1$ for some $c_1 \in \mathbf{Z}$. Then $f_1(\alpha - c_1\alpha_1) = 0$ and $f_2(\alpha - c_1\alpha_1) = c_2k_2$ for some $c_2 \in \mathbf{Z}$. Then

$$f_1(\alpha - c_1\alpha_1 - c_2\alpha_2) = f_2(\alpha - c_1\alpha_1 - c_2\alpha_2) = 0.$$

Continuing in this manner, we obtain $c_1, c_2, \ldots, c_n \in \mathbf{Z}$ such that

$$f_i(\alpha - c_1\alpha_1 - c_2\alpha_2 - \cdots - c_n\alpha_n) = 0, \qquad i = 1, 2, \ldots, n,$$

which implies that $\alpha = c_1\alpha_1 + c_2\alpha_2 + \cdots + c_n\alpha_n$.

Corollary. *The ring R is Noetherian, that is, every ideal of R is finitely generated.*

Proof. In the notation of the proposition we have that $\mathfrak{a}$ is generated by $\alpha_1, \alpha_2, \ldots, \alpha_n$.

195α. Prove that an integral domain R' which is a finite extension of a Noetherian domain is itself Noetherian.

195β. Prove that a ring is Noetherian if it is a quotient ring of a Noetherian ring.

195γ. With R as above and $n \in \mathbf{Z}$ show that $R/(n)$ is finite.

196. Theorem. *The ring of algebraic integers in a Galois extension of the rational field $\mathbf{Q}$ is a Dedekind domain.*

Proof. As above we let E be a Galois extension of $\mathbf{Q}$ of degree n with $\mathscr{G}(E/\mathbf{Q}) = \{\phi_1, \phi_2, \ldots, \phi_n\}$ and $R = \mathbf{Z}_E = \overline{\mathbf{Z}} \cap E$. It is enough to show that for any nonzero ideal $\mathfrak{a}$ of R there exists a nonzero ideal $\mathfrak{b}$ of R such that $\mathfrak{a}\mathfrak{b} = (c)$, a principal ideal; this implies that $\mathfrak{a}(\mathfrak{b}/(c)) = R$, and $\mathfrak{a}$ is invertible.

Suppose then that $\mathfrak{a}$ is a nonzero ideal of R. From **195** we know that $\mathfrak{a}$ is finitely generated, say $\mathfrak{a} = (\alpha_0, \alpha_1, \ldots, \alpha_k)$. We form the polynomial

$$fx = \alpha_0 x^k + \alpha_1 x^{k-1} + \cdots + \alpha_{k-1}x + \alpha_k$$

and let $\rho_1, \rho_2, \ldots, \rho_k \in \mathbf{C}$ denote the roots of f. Then it follows that

$$\alpha_i = (-1)^i \alpha_0 \, \sigma_i(\rho_1, \rho_2, \ldots, \rho_k),$$

where σ_i denotes the i-th symmetric function (**131**). For $\phi \in \mathscr{G}(E/\mathbf{Q})$ we let ϕf denote the polynomial given by

$$(\phi f)x = (\phi\alpha_0)x^k + (\phi\alpha_1)x^{k-1} + \cdots + (\phi\alpha_{k-1})x + (\phi\alpha_k).$$

The coefficients of ϕf are elements of R, that is, algebraic integers in E, and the roots of ϕf are $\phi\rho_1, \phi\rho_2, \ldots, \phi\rho_k$. It follows that

$$h = (\phi_1 f)(\phi_2 f) \cdots (\phi_n f)$$

is a polynomial whose coefficients are algebraic integers of E and, at the same time, rational numbers since $\phi h = h$ for $\phi \in \mathscr{G}(E/\mathbf{Q})$. Consequently the coefficients of h are rational integers. Furthermore, one element of $\mathscr{G}(E/\mathbf{Q})$ is the identity automorphism of E, and therefore f is one of the factors of h. In other words, $h = fg$ where g is a polynomial with coefficients in R. We let

$$gx = \beta_0 x^l + \beta_1 x^{l-1} + \cdots + \beta_{l-1} x + \beta_l$$

and

$$hx = \gamma_0 x^m + \gamma_1 x^{m-1} + \cdots + \gamma_{m-1} x + \gamma_m,$$

where $m = k + l$. We denote the roots of g by $\tau_1, \tau_2, \ldots, \tau_l$; then the roots of h are

$$\rho_1, \rho_2, \ldots, \rho_k, \tau_1, \tau_2, \ldots, \tau_l.$$

Since h has integral coefficients, we can write $h = c\bar{h}$ where $\bar{h}$ is a primitive polynomial and c is the content of h, which is to say, the greatest common divisor of the integers $\gamma_0, \gamma_1, \ldots, \gamma_m$. The coefficients of $\bar{h}$ are

$$(\gamma_0/c), (\gamma_1/c), \ldots, (\gamma_m/c),$$

all of which are integers, while the roots of $\bar{h}$ are the same as those of h. Therefore we know from the corollary in **193** that

$$(\gamma_0/c)\rho_{s_1}\rho_{s_2} \cdots \rho_{s_i}\tau_{r_1}\tau_{r_2} \cdots \tau_{r_j} \qquad (*)$$

is an algebraic integer for any choice of integers,

$$1 \le s_1 < s_2 < \cdots < s_i \le k, \qquad 1 \le r_1 < r_2 < \cdots < r_j \le l.$$

As a result, $\alpha_0 \beta_0 = \gamma_0$ implies

$$\begin{aligned}
\alpha_i \beta_j &= [(-1)^i \alpha_0 \sigma_i(\rho_1, \rho_2, \ldots, \rho_k)][(-1)^j \beta_0 \sigma_j(\tau_1, \tau_2, \ldots, \tau_l)] \\
&= (-1)^{i+j} \alpha_0 \beta_0 \sigma_i(\rho_1, \rho_2, \ldots, \rho_k)\sigma_j(\tau_1, \tau_2, \ldots, \tau_l) \\
&= (-1)^{i+j} c\{(\gamma_0/c)\sigma_i(\rho_1, \rho_2, \ldots, \rho_k)\sigma_j(\tau_1, \tau_2, \ldots, \tau_l)\}.
\end{aligned}$$

In the last equation the term inside the braces is the sum of all the terms of the form $(*)$; therefore it is an algebraic integer and an element of R. The result is that $\alpha_i \beta_j \in (c)$, the principal ideal of R generated by the rational integer c. (Of course this holds for any choice of i and j, $0 \le i \le k$ and $0 \le j \le l$.)

Now the coefficients of the polynomial g generate an ideal

$$\mathfrak{b} = (\beta_0, \beta_1, \ldots, \beta_l)$$

of R, and the coefficients of the polynomial h generate the principal ideal $(c) = (\gamma_0, \gamma_1, \ldots, \gamma_m)$. The product ideal $\mathfrak{ab}$ is generated by all the elements

$\alpha_i \beta_j$, each of which belongs to (c), and consequently $ab \subset (c)$. On the other hand,

$$\gamma_i = \alpha_0 \beta_i + \alpha_1 \beta_{i-1} + \cdots + \alpha_{i-1} \beta_1 + \alpha_i \beta_0$$

(if we take $\alpha_j = 0$ for $j > k$ and $\beta_j = 0$ for $j > l$). Thus the generators of (c) all belong to ab and $(c) \subset ab$. We may finally conclude that $ab = (c)$, which completes the proof.

196α. Prove that for any nonzero ideal a of R, R/a is finite.

196β. Prove that the ring of integers Z_F in any finite extension F of $\mathbf{Q}$ is a Dedekind domain. (This result is sometimes called the *fundamental theorem of algebraic number theory*. The theorem above is a special case from which the general case follows.)

197. *Norm and Trace.* Let the number field E be a Galois extension of $\mathbf{Q}$ with the group $\mathscr{G}(E/\mathbf{Q}) = \{\phi_1, \phi_2, \ldots, \phi_n\}$. To any element $\alpha \in E$ we assign two rational numbers, $N\alpha$ and $T\alpha$, called respectively the *norm* and *trace* of α. These are defined by

$$N\alpha = (\phi_1 \alpha)(\phi_2 \alpha) \cdots (\phi_n \alpha)$$

and

$$T\alpha = (\phi_1 \alpha) + (\phi_2 \alpha) + \cdots + (\phi_n \alpha).$$

We make the obvious observations about norm and trace:

(1) *Norm is multiplicative*, that is, $N(\alpha\beta) = (N\alpha)(N\beta)$.
(2) *Trace is additive*, that is, $T(\alpha + \beta) = (T\alpha) + (T\beta)$.
(3) *If α is an algebraic integer, then $N\alpha$ and $T\alpha$ are rational integers.*

197α. Show that when $[E : \mathbf{Q}] = 2$, an element $\alpha \in E$ is an algebraic integer if and only if $N\alpha$ and $T\alpha$ are rational integers.

197β. Determine the integers of $\mathbf{Q}(\sqrt{-5})$.

198. Theorem. *For p prime and $\rho = e^{2\pi i/p}$, the Kummer ring $Z(\rho)$ is a Dedekind domain.*

Proof. In view of the preceding theorem (**196**), we need to establish only that $Z(\rho)$ is the ring of integers of $\mathbf{Q}(\rho)$, which is the splitting field of $x^p - 1$ over $\mathbf{Q}$ and consequently a Galois extension of $\mathbf{Q}$.

We recall from **134** that the Galois group $\mathscr{G}(\mathbf{Q}(\rho)/\mathbf{Q})$ is isomorphic to $\mathbf{Z}_p'$ and that it consists of the $p-1$ automorphisms $\phi_1, \phi_2, \ldots, \phi_{p-1}$ determined by $\phi_k \rho = \rho^k$. We also recall that the irreducible monic polynomial for ρ over $\mathbf{Q}$ is the cyclotomic polynomial Φ_p given by

$$\Phi_p x = x^{p-1} + x^{p-2} + \cdots + x + 1 = (x - \rho)(x - \rho^2) \cdots (x - \rho^{p-1}). \qquad (*)$$

If we take $x = 1$ in $(*)$, we get $\Phi_p(1) = p = N(1 - \rho)$. This shows among other things that $1 - \rho$ is prime in $\mathbf{Z}_{\mathbf{Q}(\rho)}$, the ring of algebraic integers of $\mathbf{Q}(\rho)$.

Next we remark that $(1 - \rho) \cap \mathbf{Z} = (p)$ in $\mathbf{Z}_{\mathbf{Q}(\rho)}$ or, in other words, *a rational integer divisible by $(1 - \rho)$ is divisible by p.* This is easy to see since $(1 - \rho) \cap \mathbf{Z}$ is a proper ideal of $\mathbf{Z}$ and contains (p), which is a maximal ideal.

Let α be an algebraic integer of $\mathbf{Q}(\rho)$, that is, an element of $\mathbf{Z}_{\mathbf{Q}(\rho)}$. Since the numbers $\rho, \rho^2, \ldots, \rho^{p-1}$ form a basis for $\mathbf{Q}(\rho)$ over $\mathbf{Q}$, we can write α uniquely in the form

$$\alpha = a_1 \rho + a_2 \rho^2 + \cdots + a_{p-1} \rho^{p-1},$$

where $a_1, a_2, \ldots, a_{p-1} \in \mathbf{Q}$. We will have $\alpha \in \mathbf{Z}(\rho)$ if we can show that $a_1, a_2, \ldots, a_{p-1} \in \mathbf{Z}$. First we remark that

$$1 + T(\rho^k) = 1 + \rho^k + \rho^{2k} + \cdots + \rho^{(p-1)k} = \Phi_p(\rho^k) = \begin{cases} p & \text{if } p \mid k, \\ 0 & \text{if } p \nmid k. \end{cases}$$

Therefore $T(1) = p - 1$ and $T(\rho^k) = -1$ for $k = 1, 2, \ldots, p - 1$. Next we compute the trace of $(1 - \rho)\rho^{-i}\alpha$ for $i = 1, 2, \ldots, p - 1$. We have

$$\begin{aligned} T[(1 - \rho)\rho^{-i}\alpha] &= T\left[(1 - \rho)\rho^{-i}\sum_{j=1}^{p-1} a_j \rho^j\right] \\ &= \sum_{j=1}^{p-1} a_j T[(1 - \rho)\rho^{j-i}] \\ &= \sum_{j=1}^{p-1} a_j [T(\rho^{j-i}) - T(\rho^{j-i+1})] \\ &= \begin{cases} pa_1 & \text{if } i = 1, \\ p(a_i - a_{i-1}) & \text{if } i = 2, 3, \ldots, p - 1. \end{cases} \end{aligned}$$

However, $(1 - \rho)\rho^{-i}$ is an algebraic integer divisible by $1 - \rho$, and consequently its trace is a rational integer divisible by $1 - \rho$ (why?) and, hence, also divisible by p. As a result all the numbers $a_1, a_2 - a_1, \ldots, a_{p-1} - a_{p-2}$ are rational integers, from which it follows that $\alpha \in \mathbf{Z}(\rho)$. Thus we have shown that $\mathbf{Z}(\rho) \supset \mathbf{Z}_{\mathbf{Q}(\rho)}$. Therefore $\mathbf{Z}(\rho) = \mathbf{Z}_{\mathbf{Q}(\rho)}$ and we are finished.

198α.   Show that in $\mathbf{Z}(\rho)$ the ideal (p) has the factorization $(1 - \rho)^{p-1}$.

198β.   Let q be a rational prime. When is (q) a prime ideal of $\mathbf{Z}(\rho)$?

Bibliography

eneral References

Artin, Emil, *Galois Theory*, second edition, Notre Dame Mathematical Lectures, No. 2.

Birkhoff, G., and S. MacLane, *A Survey of Modern Algebra*, revised edition. New York: The Macmillan Company, 1953.

Eves, Howard, *A Survey of Geometry*, vol. I. Boston: Allyn and Bacon, Inc., 1963.

Hall, Marshall, *Theory of Groups*. New York: The Macmillan Company, 1959.

Hardy, G. H., and E. M. Wright, *An Introduction to the Theory of Numbers*, fourth edition. Oxford: The Clarendon Press, 1960.

Postnikov, M. M., *Fundamentals of Galois Theory*. Groningen: P. Noordhoff, Ltd., 1962.

van der Waerden, B. L., *Modern Algebra*. New York: F. Ungar Publishing Company, 1949.

Zariski, O., and P. Samuel, *Commutative Algebra*, vol. I. Princeton: D. Van Nostrand Company, 1958.

Historical References

---

al-Khwārizmī, *Robert of Chester's Latin Translation of the Algebra of Muhammed ben Musa.* New York: The Macmillan Company, 1915. Contains an English translation of the Latin version.

Burkhardt, H., "Endliche Diskrete Gruppen," *Encyclopädie der Mathematischen Wissenschaften*, Band I, Teil I, Heft 3. Leipzig, 1899. A survey of the history of the theory of finite groups up to 1899. (In German.)

Cayley, Arthur, "On the theory of groups as depending on the symbolical equation $\theta^n = 1$," *Collected Works*, vol. II, pp. 123–132. Cambridge: The University Press, 1889–97. Two short, easy articles in which groups are discussed abstractly.

Cardano, Girolamo, *The Great Art or the Rules of Algebra.* Cambridge, Massachusetts: The M.I.T. Press, 1968.

Dedekind, Richard, *Sur la Théorie des Nombres Entiers Algébriques.* Paris, 1877. A beautiful little introduction to algebraic integers and ideal theory. (In French.)

Euler, Leonard, *An Introduction to the Elements of Algebra*, fourth edition. Boston: Hilliard, Gray, and Company, 1836. A classic elementary textbook.

Galois, Évariste, *Écrits et mémoirs mathématiques.* Paris: Gauthier-Villars, 1962. The definitive edition of the complete works of Galois. (In French.)

Gauss, Karl Friedrich, *Disquisitiones Arithmeticae.* New Haven: Yale University Press, 1966. (In English.)

Lagrange, Joseph Louis, "Réflexions sur la Résolution Algébrique des Equations," *Oeuvres de Lagrange*, vol. 3, pp. 205–421. Paris: Gauthier-Villars, 1869. (In French.)

Ruffini, Paolo, *Teoria generale della equazioni in ciu si dimostra impossible la soluzione algebraica della equazioni generali di grado superiore al quarto.* Bologna, 1799. (In Italian.)

Waring, Edward, *Meditationes Algebraicae.* Cantabrigiae, 1770. (In Latin.)

Index

The Greek Alphabet

A	α	*alpha*		N	ν	*nu*
B	β	*beta*		Ξ	ξ	*xi*
Γ	γ	*gamma*		O	o	*omicron*
Δ	δ	*delta*		Π	π	*pi*
E	ε	*epsilon*		P	ρ	*rho*
Z	ζ	*zeta*		Σ	σ	*sigma*
H	η	*eta*		T	τ	*tau*
Θ	θ	*theta*		Υ	υ	*upsilon*
I	ι	*iota*		Φ	ϕ	*phi*
K	κ	*kappa*		X	χ	*chi*
Λ	λ	*lambda*		Ψ	ψ	*psi*
M	μ	*mu*		Ω	ω	*omega*

Symbols for Special Sets

N	*natural numbers*
$\mathbf{N}_k$	$\{1, 2, \ldots, k\}$
Z	*integers*
$\mathbf{Z}_n$	*integers modulo n*
$\mathbf{Z}'_n$	*units of* $\mathbf{Z}_n$
Z(i)	*Gaussian integers*
Q	*rational numbers*
R	*real numbers*
C	*complex numbers*

A CATALOGUE OF
SELECTED DOVER BOOKS
IN ALL FIELDS OF INTEREST

A CATALOGUE OF SELECTED DOVER
BOOKS IN ALL FIELDS OF INTEREST

CELESTIAL OBJECTS FOR COMMON TELESCOPES, T. W. Webb. The most used book in amateur astronomy: inestimable aid for locating and identifying nearly 4,000 celestial objects. Edited, updated by Margaret W. Mayall. 77 illustrations. Total of 645pp. 5⅜ x 8½.
20917-2, 20918-0 Pa., Two-vol. set $10.00

HISTORICAL STUDIES IN THE LANGUAGE OF CHEMISTRY, M. P. Crosland. The important part language has played in the development of chemistry from the symbolism of alchemy to the adoption of systematic nomenclature in 1892. ". . . wholeheartedly recommended,"—Science. 15 illustrations. 416pp. of text. 5⅝ x 8¼. 63702-6 Pa. $7.50

BURNHAM'S CELESTIAL HANDBOOK, Robert Burnham, Jr. Thorough, readable guide to the stars beyond our solar system. Exhaustive treatment, fully illustrated. Breakdown is alphabetical by constellation: Andromeda to Cetus in Vol. 1; Chamaeleon to Orion in Vol. 2; and Pavo to Vulpecula in Vol. 3. Hundreds of illustrations. Total of about 2000pp. 6⅛ x 9¼.
23567-X, 23568-8, 23673-0 Pa., Three-vol. set $32.85

THEORY OF WING SECTIONS: INCLUDING A SUMMARY OF AIR-FOIL DATA, Ira H. Abbott and A. E. von Doenhoff. Concise compilation of subatomic aerodynamic characteristics of modern NASA wing sections, plus description of theory. 350pp. of tables. 693pp. 5⅜ x 8½.
60586-8 Pa. $9.95

DE RE METALLICA, Georgius Agricola. Translated by Herbert C. Hoover and Lou H. Hoover. The famous Hoover translation of greatest treatise on technological chemistry, engineering, geology, mining of early modern times (1556). All 289 original woodcuts. 638pp. 6¾ x 11.
60006-8 Clothbd. $19.95

THE ORIGIN OF CONTINENTS AND OCEANS, Alfred Wegener. One of the most influential, most controversial books in science, the classic statement for continental drift. Full 1966 translation of Wegener's final (1929) version. 64 illustrations. 246pp. 5⅜ x 8½.(EBE)61708-4 Pa. $5.00

THE PRINCIPLES OF PSYCHOLOGY, William James. Famous long course complete, unabridged. Stream of thought, time perception, memory, experimental methods; great work decades ahead of its time. Still valid, useful; read in many classes. 94 figures. Total of 1391pp. 5⅜ x 8½.
20381-6, 20382-4 Pa., Two-vol. set $17.90

YUCATAN BEFORE AND AFTER THE CONQUEST, Diego de Landa. First English translation of basic book in Maya studies, the only significant account of Yucatan written in the early post-Conquest era. Translated by distinguished Maya scholar William Gates. Appendices, introduction, 4 maps and over 120 illustrations added by translator. 162pp. 5⅜ x 8½.
23622-6 Pa. $3.00

THE MALAY ARCHIPELAGO, Alfred R. Wallace. Spirited travel account by one of founders of modern biology. Touches on zoology, botany, ethnography, geography, and geology. 62 illustrations, maps. 515pp. 5⅜ x 8½.
20187-2 Pa. $6.95

THE DISCOVERY OF THE TOMB OF TUTANKHAMEN, Howard Carter, A. C. Mace. Accompany Carter in the thrill of discovery, as ruined passage suddenly reveals unique, untouched, fabulously rich tomb. Fascinating account, with 106 illustrations. New introduction by J. M. White. Total of 382pp. 5⅜ x 8½. (Available in U.S. only) 23500-9 Pa. $5.50

THE WORLD'S GREATEST SPEECHES, edited by Lewis Copeland and Lawrence W. Lamm. Vast collection of 278 speeches from Greeks up to present. Powerful and effective models; unique look at history. Revised to 1970. Indices. 842pp. 5⅜ x 8½. 20468-5 Pa. $9.95

THE 100 GREATEST ADVERTISEMENTS, Julian Watkins. The priceless ingredient; His master's voice; 99 44/100% pure; over 100 others. How they were written, their impact, etc. Remarkable record. 130 illustrations. 233pp. 7⅞ x 10 3/5. 20540-1 Pa. $6.95

CRUICKSHANK PRINTS FOR HAND COLORING, George Cruickshank. 18 illustrations, one side of a page, on fine-quality paper suitable for watercolors. Caricatures of people in society (c. 1820) full of trenchant wit. Very large format. 32pp. 11 x 16. 23684-6 Pa. $6.00

THIRTY-TWO COLOR POSTCARDS OF TWENTIETH-CENTURY AMERICAN ART, Whitney Museum of American Art. Reproduced in full color in postcard form are 31 art works and one shot of the museum. Calder, Hopper, Rauschenberg, others. Detachable. 16pp. 8¼ x 11.
23629-3 Pa. $3.50

MUSIC OF THE SPHERES: THE MATERIAL UNIVERSE FROM ATOM TO QUASAR SIMPLY EXPLAINED, Guy Murchie. Planets, stars, geology, atoms, radiation, relativity, quantum theory, light, antimatter, similar topics. 319 figures. 664pp. 5⅜ x 8½.
21809-0, 21810-4 Pa., Two-vol. set $11.00

EINSTEIN'S THEORY OF RELATIVITY, Max Born. Finest semi-technical account; covers Einstein, Lorentz, Minkowski, and others, with much detail, much explanation of ideas and math not readily available elsewhere on this level. For student, non-specialist. 376pp. 5⅜ x 8½.
60769-0 Pa. $5.00

THE SENSE OF BEAUTY, George Santayana. Masterfully written discussion of nature of beauty, materials of beauty, form, expression; art, literature, social sciences all involved. 168pp. 5⅜ x 8½.   20238-0 Pa. $3.50

ON THE IMPROVEMENT OF THE UNDERSTANDING, Benedict Spinoza. Also contains *Ethics, Correspondence,* all in excellent R. Elwes translation. Basic works on entry to philosophy, pantheism, exchange of ideas with great contemporaries. 402pp. 5⅜ x 8½.   20250-X Pa. $5.95

THE TRAGIC SENSE OF LIFE, Miguel de Unamuno. Acknowledged masterpiece of existential literature, one of most important books of 20th century. Introduction by Madariaga. 367pp. 5⅜ x 8½.
20257-7 Pa. $6.00

THE GUIDE FOR THE PERPLEXED, Moses Maimonides. Great classic of medieval Judaism attempts to reconcile revealed religion (Pentateuch, commentaries) with Aristotelian philosophy. Important historically, still relevant in problems. Unabridged Friedlander translation. Total of 473pp. 5⅜ x 8½.   20351-4 Pa. $6.95

THE I CHING (THE BOOK OF CHANGES), translated by James Legge. Complete translation of basic text plus appendices by Confucius, and Chinese commentary of most penetrating divination manual ever prepared. Indispensable to study of early Oriental civilizations, to modern inquiring reader. 448pp. 5⅜ x 8½.   21062-6 Pa. $6.00

THE EGYPTIAN BOOK OF THE DEAD, E. A. Wallis Budge. Complete reproduction of Ani's papyrus, finest ever found. Full hieroglyphic text, interlinear transliteration, word for word translation, smooth translation. Basic work, for Egyptology, for modern study of psychic matters. Total of 533pp. 6½ x 9¼. (USCO)   21866-X Pa. $8.50

THE GODS OF THE EGYPTIANS, E. A. Wallis Budge. Never excelled for richness, fullness: all gods, goddesses, demons, mythical figures of Ancient Egypt; their legends, rites, incarnations, variations, powers, etc. Many hieroglyphic texts cited. Over 225 illustrations, plus 6 color plates. Total of 988pp. 6⅛ x 9¼. (EBE)
22055-9, 22056-7 Pa., Two-vol. set $20.00

THE STANDARD BOOK OF QUILT MAKING AND COLLECTING, Marguerite Ickis. Full information, full-sized patterns for making 46 traditional quilts, also 150 other patterns. Quilted cloths, lame, satin quilts, etc. 483 illustrations. 273pp. 6⅞ x 9⅝.   20582-7 Pa. $5.95

CORAL GARDENS AND THEIR MAGIC, Bronsilaw Malinowski. Classic study of the methods of tilling the soil and of agricultural rites in the Trobriand Islands of Melanesia. Author is one of the most important figures in the field of modern social anthropology. 143 illustrations. Indexes. Total of 911pp. of text. 5⅝ x 8¼. (Available in U.S. only)
23597-1 Pa. $12.95

THE PHILOSOPHY OF HISTORY, Georg W. Hegel. Great classic of Western thought develops concept that history is not chance but a rational process, the evolution of freedom. 457pp. 5⅜ x 8½. 20112-0 Pa. $6.00

LANGUAGE, TRUTH AND LOGIC, Alfred J. Ayer. Famous, clear introduction to Vienna, Cambridge schools of Logical Positivism. Role of philosophy, elimination of metaphysics, nature of analysis, etc. 160pp. 5⅜ x 8½. (USCO) 20010-8 Pa. $2.50

A PREFACE TO LOGIC, Morris R. Cohen. Great City College teacher in renowned, easily followed exposition of formal logic, probability, values, logic and world order and similar topics; no previous background needed. 209pp. 5⅜ x 8½. 23517-3 Pa. $4.95

REASON AND NATURE, Morris R. Cohen. Brilliant analysis of reason and its multitudinous ramifications by charismatic teacher. Interdisciplinary, synthesizing work widely praised when it first appeared in 1931. Second (1953) edition. Indexes. 496pp. 5⅜ x 8½. 23633-1 Pa. $7.50

AN ESSAY CONCERNING HUMAN UNDERSTANDING, John Locke. The only complete edition of enormously important classic, with authoritative editorial material by A. C. Fraser. Total of 1176pp. 5⅜ x 8½.
20530-4, 20531-2 Pa., Two-vol. set $16.00

HANDBOOK OF MATHEMATICAL FUNCTIONS WITH FORMULAS, GRAPHS, AND MATHEMATICAL TABLES, edited by Milton Abramowitz and Irene A. Stegun. Vast compendium: 29 sets of tables, some to as high as 20 places. 1,046pp. 8 x 10½. 61272-4 Pa. $17.95

MATHEMATICS FOR THE PHYSICAL SCIENCES, Herbert S. Wilf. Highly acclaimed work offers clear presentations of vector spaces and matrices, orthogonal functions, roots of polynomial equations, conformal mapping, calculus of variations, etc. Knowledge of theory of. functions of real and complex variables is assumed. Exercises and solutions. Index. 284pp. 5⅝ x 8¼. 63635-6 Pa. $5.00

THE PRINCIPLE OF RELATIVITY, Albert Einstein et al. Eleven most important original papers on special and general theories. Seven by Einstein, two by Lorentz, one each by Minkowski and Weyl. All translated, unabridged. 216pp. 5⅜ x 8½. 60081-5 Pa. $3.50

THERMODYNAMICS, Enrico Fermi. A classic of modern science. Clear, organized treatment of systems, first and second laws, entropy, thermodynamic potentials, gaseous reactions, dilute solutions, entropy constant. No math beyond calculus required. Problems. 160pp. 5⅜ x 8½.
60361-X Pa. $4.00

ELEMENTARY MECHANICS OF FLUIDS, Hunter Rouse. Classic undergraduate text widely considered to be far better than many later books. Ranges from fluid velocity and acceleration to role of compressibility in fluid motion. Numerous examples, questions, problems. 224 illustrations. 376pp. 5⅝ x 8¼. 63699-2 Pa. $7.00

THE AMERICAN SENATOR, Anthony Trollope. Little known, long unavailable Trollope novel on a grand scale. Here are humorous comment on American vs. English culture, and stunning portrayal of a heroine/villainess. Superb evocation of Victorian village life. 561pp. 5⅜ x 8½.
23801-6 Pa. $7.95

WAS IT MURDER? James Hilton. The author of *Lost Horizon* and *Goodbye, Mr. Chips* wrote one detective novel (under a pen-name) which was quickly forgotten and virtually lost, even at the height of Hilton's fame. This edition brings it back—a finely crafted public school puzzle resplendent with Hilton's stylish atmosphere. A thoroughly English thriller by the creator of Shangri-la. 252pp. 5⅜ x 8. (Available in U.S. only)
23774-5 Pa. $3.00

CENTRAL PARK: A PHOTOGRAPHIC GUIDE, Victor Laredo and Henry Hope Reed. 121 superb photographs show dramatic views of Central Park: Bethesda Fountain, Cleopatra's Needle, Sheep Meadow, the Blockhouse, plus people engaged in many park activities: ice skating, bike riding, etc. Captions by former Curator of Central Park, Henry Hope Reed, provide historical view, changes, etc. Also photos of N.Y. landmarks on park's periphery. 96pp. 8½ x 11. 23750-8 Pa. $4.50

NANTUCKET IN THE NINETEENTH CENTURY, Clay Lancaster. 180 rare photographs, stereographs, maps, drawings and floor plans recreate unique American island society. Authentic scenes of shipwreck, lighthouses, streets, homes are arranged in geographic sequence to provide walking-tour guide to old Nantucket existing today. Introduction, captions. 160pp. 8⅞ x 11¾. 23747-8 Pa. $7.95

STONE AND MAN: A PHOTOGRAPHIC EXPLORATION, Andreas Feininger. 106 photographs by *Life* photographer Feininger portray man's deep passion for stone through the ages. Stonehenge-like megaliths, fortified towns, sculpted marble and crumbling tenements show textures, beauties, fascination. 128pp. 9¼ x 10¾. 23756-7 Pa. $5.95

CIRCLES, A MATHEMATICAL VIEW, D. Pedoe. Fundamental aspects of college geometry, non-Euclidean geometry, and other branches of mathematics: representing circle by point. Poincare model, isoperimetric property, etc. Stimulating recreational reading. 66 figures. 96pp. 5⅝ x 8¼.
63698-4 Pa. $3.50

THE DISCOVERY OF NEPTUNE, Morton Grosser. Dramatic scientific history of the investigations leading up to the actual discovery of the eighth planet of our solar system. Lucid, well-researched book by well-known historian of science. 172pp. 5⅜ x 8½. 23726-5 Pa. $3.50

THE DEVIL'S DICTIONARY. Ambrose Bierce. Barbed, bitter, brilliant witticisms in the form of a dictionary. Best, most ferocious satire America has produced. 145pp. 5⅜ x 8½. 20487-1 Pa. $2.50

HISTORY OF BACTERIOLOGY, William Bulloch. The only comprehensive history of bacteriology from the beginnings through the 19th century. Special emphasis is given to biography-Leeuwenhoek, etc. Brief accounts of 350 bacteriologists form a separate section. No clearer, fuller study, suitable to scientists and general readers, has yet been written. 52 illustrations. 448pp. 5⅝ x 8¼. 23761-3 Pa. $6.50

THE COMPLETE NONSENSE OF EDWARD LEAR, Edward Lear. All nonsense limericks, zany alphabets, Owl and Pussycat, songs, nonsense botany, etc., illustrated by Lear. Total of 321pp. 5⅜ x 8½. (Available in U.S. only) 20167-8 Pa. $4.50

INGENIOUS MATHEMATICAL PROBLEMS AND METHODS, Louis A. Graham. Sophisticated material from Graham Dial, applied and pure; stresses solution methods. Logic, number theory, networks, inversions, etc. 237pp. 5⅜ x 8½. 20545-2 Pa. $4.50

BEST MATHEMATICAL PUZZLES OF SAM LOYD, edited by Martin Gardner. Bizarre, original, whimsical puzzles by America's greatest puzzler. From fabulously rare Cyclopedia, including famous 14-15 puzzles, the Horse of a Different Color, 115 more. Elementary math. 150 illustrations. 167pp. 5⅜ x 8½. 20498-7 Pa. $3.50

THE BASIS OF COMBINATION IN CHESS, J. du Mont. Easy-to-follow, instructive book on elements of combination play, with chapters on each piece and every powerful combination team—two knights, bishop and knight, rook and bishop, etc. 250 diagrams. 218pp. 5⅜ x 8½. (Available in U.S. only) 23644-7 Pa. $4.50

MODERN CHESS STRATEGY, Ludek Pachman. The use of the queen, the active king, exchanges, pawn play, the center, weak squares, etc. Section on rook alone worth price of the book. Stress on the moderns. Often considered the most important book on strategy. 314pp. 5⅜ x 8½. 20290-9 Pa. $5.00

LASKER'S MANUAL OF CHESS, Dr. Emanuel Lasker. Great world champion offers very thorough coverage of all aspects of chess. Combinations, position play, openings, end game, aesthetics of chess, philosophy of struggle, much more. Filled with analyzed games. 390pp. 5⅜ x 8½. 20640-8 Pa. $5.95

500 MASTER GAMES OF CHESS, S. Tartakower, J. du Mont. Vast collection of great chess games from 1798-1938, with much material nowhere else readily available. Fully annotated, arranged by opening for easier study. 664pp. 5⅜ x 8½. 23208-5 Pa. $8.50

A GUIDE TO CHESS ENDINGS, Dr. Max Euwe, David Hooper. One of the finest modern works on chess endings. Thorough analysis of the most frequently encountered endings by former world champion. 331 examples, each with diagram. 248pp. 5⅜ x 8½. 23332-4 Pa. $3.95

THE COMPLETE BOOK OF DOLL MAKING AND COLLECTING, Catherine Christopher. Instructions, patterns for dozens of dolls, from rag doll on up to elaborate, historically accurate figures. Mould faces, sew clothing, make doll houses, etc. Also collecting information. Many illustrations. 288pp. 6 x 9. 22066-4 Pa. $4.95

THE DAGUERREOTYPE IN AMERICA, Beaumont Newhall. Wonderful portraits, 1850's townscapes, landscapes; full text plus 104 photographs. The basic book. Enlarged 1976 edition. 272pp. 8¼ x 11¼. 23322-7 Pa. $7.95

CRAFTSMAN HOMES, Gustav Stickley. 296 architectural drawings, floor plans, and photographs illustrate 40 different kinds of "Mission-style" homes from The Craftsman (1901-16), voice of American style of simplicity and organic harmony. Thorough coverage of Craftsman idea in text and picture, now collector's item. 224pp. 8⅛ x 11. 23791-5 Pa. $6.50

PEWTER-WORKING: INSTRUCTIONS AND PROJECTS, Burl N. Osborn. & Gordon O. Wilber. Introduction to pewter-working for amateur craftsman. History and characteristics of pewter; tools, materials, step-by-step instructions. Photos, line drawings, diagrams. Total of 160pp. 7⅞ x 10¾. 23786-9 Pa. $3.50

THE GREAT CHICAGO FIRE, edited by David Lowe. 10 dramatic, eye-witness accounts of the 1871 disaster, including one of the aftermath and rebuilding, plus 70 contemporary photographs and illustrations of the ruins—courthouse, Palmer House, Great Central Depot, etc. Introduction by David Lowe. 87pp. 8¼ x 11. 23771-0 Pa. $4.00

SILHOUETTES: A PICTORIAL ARCHIVE OF VARIED ILLUSTRA-TIONS, edited by Carol Belanger Grafton. Over 600 silhouettes from the 18th to 20th centuries include profiles and full figures of men and women, children, birds and animals, groups and scenes, nature, ships, an alphabet. Dozens of uses for commercial artists and craftspeople. 144pp. 8⅜ x 11¼. 23781-8 Pa. $4.50

ANIMALS: 1,419 COPYRIGHT-FREE ILLUSTRATIONS OF MAM-MALS, BIRDS, FISH, INSECTS, ETC., edited by Jim Harter. Clear wood engravings present, in extremely lifelike poses, over 1,000 species of animals. One of the most extensive copyright-free pictorial sourcebooks of its kind. Captions. Index. 284pp. 9 x 12. 23766-4 Pa. $8.95

INDIAN DESIGNS FROM ANCIENT ECUADOR, Frederick W. Shaffer. 282 original designs by pre-Columbian Indians of Ecuador (500-1500 A.D.). Designs include people, mammals, birds, reptiles, fish, plants, heads, geo-metric designs. Use as is or alter for advertising, textiles, leathercraft, etc. Introduction. 95pp. 8¾ x 11¼. 23764-8 Pa. $4.50

SZIGETI ON THE VIOLIN, Joseph Szigeti. Genial, loosely structured tour by premier violinist, featuring a pleasant mixture of reminiscenes, insights into great music and musicians, innumerable tips for practicing violinists. 385 musical passages. 256pp. 5⅝ x 8¼. 23763-X Pa. $4.00

TONE POEMS, SERIES II: TILL EULENSPIEGELS LUSTIGE STREICHE, ALSO SPRACH ZARATHUSTRA, AND EIN HELDEN-LEBEN, Richard Strauss. Three important orchestral works, including very popular *Till Eulenspiegel's Marry Pranks,* reproduced in full score from original editions. Study score. 315pp. 9⅜ x 12¼. (Available in U.S. only)
23755-9 Pa. $8.95

TONE POEMS, SERIES I: DON JUAN, TOD UND VERKLARUNG AND DON QUIXOTE, Richard Strauss. Three of the most often performed and recorded works in entire orchestral repertoire, reproduced in full score from original editions. Study score. 286pp. 9⅜ x 12¼. (Available in U.S. only)
23754-0 Pa. $8.95

11 LATE STRING QUARTETS, Franz Joseph Haydn. The form which Haydn defined and "brought to perfection." (*Grove's*). 11 string quartets in complete score, his last and his best. The first in a projected series of the complete Haydn string quartets. Reliable modern Eulenberg edition, otherwise difficult to obtain. 320pp. 8⅜ x 11¼. (Available in U.S. only)
23753-2 Pa. $8.95

FOURTH, FIFTH AND SIXTH SYMPHONIES IN FULL SCORE, Peter Ilyitch Tchaikovsky. Complete orchestral scores of Symphony No. 4 in F Minor, Op. 36; Symphony No. 5 in E Minor, Op. 64; Symphony No. 6 in B Minor, "Pathetique," Op. 74. Bretikopf & Hartel eds. Study score. 480pp. 9⅜ x 12¼.
23861-X Pa. $10.95

THE MARRIAGE OF FIGARO: COMPLETE SCORE, Wolfgang A. Mozart. Finest comic opera ever written. Full score, not to be confused with piano renderings. Peters edition. Study score. 448pp. 9⅜ x 12¼. (Available in U.S. only)
23751-6 Pa. $12.95

"IMAGE" ON THE ART AND EVOLUTION OF THE FILM, edited by Marshall Deutelbaum. Pioneering book brings together for first time 38 groundbreaking articles on early silent films from *Image* and 263 illustrations newly shot from rare prints in the collection of the International Museum of Photography. A landmark work. Index. 256pp. 8¼ x 11.
23777-X Pa. $8.95

AROUND-THE-WORLD COOKY BOOK, Lois Lintner Sumption and Marguerite Lintner Ashbrook. 373 cooky and frosting recipes from 28 countries (America, Austria, China, Russia, Italy, etc.) include Viennese kisses, rice wafers, London strips, lady fingers, hony, sugar spice, maple cookies, etc. Clear instructions. All tested. 38 drawings. 182pp. 5⅜ x 8.
23802-4 Pa. $2.75

THE ART NOUVEAU STYLE, edited by Roberta Waddell. 579 rare photographs, not available elsewhere, of works in jewelry, metalwork, glass, ceramics, textiles, architecture and furniture by 175 artists—Mucha, Seguy, Lalique, Tiffany, Gaudin, Hohlwein, Saarinen, and many others. 288pp. 8⅜ x 11¼.
23515-7 Pa. $8.95

THE CURVES OF LIFE, Theodore A. Cook. Examination of shells, leaves, horns, human body, art, etc., in *"the* classic reference on how the golden ratio applies to spirals and helices in nature . . . . "—Martin Gardner. 426 illustrations. Total of 512pp. 5⅜ x 8½.   23701-X Pa. **$6.95**

AN ILLUSTRATED FLORA OF THE NORTHERN UNITED STATES AND CANADA, Nathaniel L. Britton, Addison Brown. Encyclopedic work covers 4666 species, ferns on up. Everything. Full botanical information, illustration for each. This earlier edition is preferred by many to more recent revisions. 1913 edition. Over 4000 illustrations, total of 2087pp. 6⅛ x 9¼.   22642-5, 22643-3, 22644-1 Pa., Three-vol. set **$28.50**

MANUAL OF THE GRASSES OF THE UNITED STATES, A. S. Hitchcock, U.S. Dept. of Agriculture. The basic study of American grasses, both indigenous and escapes, cultivated and wild. Over 1400 species. Full descriptions, information. Over 1100 maps, illustrations. Total of 1051pp. 5⅜ x 8½.   22717-0, 22718-9 Pa., Two-vol. set **$17.00**

THE CACTACEAE,, Nathaniel L. Britton, John N. Rose. Exhaustive, definitive. Every cactus in the world. Full botanical descriptions. Thorough statement of nomenclatures, habitat, detailed finding keys. The one book needed by every cactus enthusiast. Over 1275 illustrations. Total of 1080pp. 8 x 10¼.   21191-6, 21192-4 Clothbd., Two-vol. set **$50.00**

AMERICAN MEDICINAL PLANTS, Charles F. Millspaugh. Full descriptions, 180 plants covered: history; physical description; methods of preparation with all chemical constituents extracted; all claimed curative or adverse effects. 180 full-page plates. Classification table. 804pp. 6½ x 9¼.
23034-1 Pa. **$13.95**

A MODERN HERBAL, Margaret Grieve. Much the fullest, most exact, most useful compilation of herbal material. Gigantic alphabetical encyclopedia, from aconite to zedoary, gives botanical information, medical properties, folklore, economic uses, and much else. Indispensable to serious reader. 161 illustrations. 888pp. 6½ x 9¼. (Available in U.S. only)
22798-7, 22799-5 Pa., Two-vol. set **$15.00**

THE HERBAL or GENERAL HISTORY OF PLANTS, John Gerard. The 1633 edition revised and enlarged by Thomas Johnson. Containing almost 2850 plant descriptions and 2705 superb illustrations, Gerard's *Herbal* is a monumental work, the book all modern English herbals are derived from, the one herbal every serious enthusiast should have in its entirety. Original editions are worth perhaps $750. 1678pp. 8½ x 12¼.
23147-X Clothbd. **$75.00**

MANUAL OF THE TREES OF NORTH AMERICA, Charles S. Sargent. The basic survey of every native tree and tree-like shrub, 717 species in all. Extremely full descriptions, information on habitat, growth, locales, economics, etc. Necessary to every serious tree lover. Over 100 finding keys. 783 illustrations. Total of 986pp. 5⅜ x 8½.
20277-1, 20278-X Pa., Two-vol. set **$12.00**

GREAT NEWS PHOTOS AND THE STORIES BEHIND THEM, John Faber. Dramatic volume of 140 great news photos, 1855 through 1976, and revealing stories behind them, with both historical and technical information. Hindenburg disaster, shooting of Oswald, nomination of Jimmy Carter, etc. 160pp. 8¼ x 11. 23667-6 Pa. $6.00

CRUICKSHANK'S PHOTOGRAPHS OF BIRDS OF AMERICA, Allan D. Cruickshank. Great ornithologist, photographer presents 177 closeups, groupings, panoramas, flightings, etc., of about 150 different birds. Expanded *Wings in the Wilderness*. Introduction by Helen G. Cruickshank. 191pp. 8¼ x 11. 23497-5 Pa. $7.95

AMERICAN WILDLIFE AND PLANTS, A. C. Martin, et al. Describes food habits of more than 1000 species of mammals, birds, fish. Special treatment of important food plants. Over 300 illustrations. 500pp. 5⅜ x 8½. 20793-5 Pa. $6.50

THE PEOPLE CALLED SHAKERS, Edward D. Andrews. Lifetime of research, definitive study of Shakers: origins, beliefs, practices, dances, social organization, furniture and crafts, impact on 19th-century USA, present heritage. Indispensable to student of American history, collector. 33 illustrations. 351pp. 5⅜ x 8½. 21081-2 Pa. $4.50

OLD NEW YORK IN EARLY PHOTOGRAPHS, Mary Black. New York City as it was in 1853-1901, through 196 wonderful photographs from N.-Y. Historical Society. Great Blizzard, Lincoln's funeral procession, great buildings. 228pp. 9 x 12. 22907-6 Pa. $8.95

MR. LINCOLN'S CAMERA MAN: MATHEW BRADY, Roy Meredith. Over 300 Brady photos reproduced directly from original negatives, photos. Jackson, Webster, Grant, Lee, Carnegie, Barnum; Lincoln; Battle Smoke, Death of Rebel Sniper, Atlanta Just After Capture. Lively commentary. 368pp. 8⅜ x 11¼. 23021-X Pa. $11.95

TRAVELS OF WILLIAM BARTRAM, William Bartram. From 1773-8, Bartram explored Northern Florida, Georgia, Carolinas, and reported on wild life, plants, Indians, early settlers. Basic account for period, entertaining reading. Edited by Mark Van Doren. 13 illustrations. 141pp. 5⅜ x 8½. 20013-2 Pa. $6.00

THE GENTLEMAN AND CABINET MAKER'S DIRECTOR, Thomas Chippendale. Full reprint, 1762 style book, most influential of all time; chairs, tables, sofas, mirrors, cabinets, etc. 200 plates, plus 24 photographs of surviving pieces. 249pp. 9⅞ x 12¾. 21601-2 Pa. $8.95

AMERICAN CARRIAGES, SLEIGHS, SULKIES AND CARTS, edited by Don H. Berkebile. 168 Victorian illustrations from catalogues, trade journals, fully captioned. Useful for artists. Author is Assoc. Curator, Div. of Transportation of Smithsonian Institution. 168pp. 8½ x 9½. 23328-6 Pa. $5.00

SECOND PIATIGORSKY CUP, edited by Isaac Kashdan. One of the greatest tournament books ever produced in the English language. All 90 games of the 1966 tournament, annotated by players, most annotated by both players. Features Petrosian, Spassky, Fischer, Larsen, six others. 228pp. 5⅜ x 8½.   23572-6 Pa. $3.50

ENCYCLOPEDIA OF CARD TRICKS, revised and edited by Jean Hugard. How to perform over 600 card tricks, devised by the world's greatest magicians: impromptus, spelling tricks, key cards, using special packs, much, much more. Additional chapter on card technique. 66 illustrations. 402pp. 5⅜ x 8½. (Available in U.S. only)   21252-1 Pa. $5.95

MAGIC: STAGE ILLUSIONS, SPECIAL EFFECTS AND TRICK PHO-TOGRAPHY, Albert A. Hopkins, Henry R. Evans. One of the great classics; fullest, most authorative explanation of vanishing lady, levitations, scores of other great stage effects. Also small magic, automata, stunts. 446 illus-trations. 556pp. 5⅜ x 8½.   23344-8 Pa. $6.95

THE SECRETS OF HOUDINI, J. C. Cannell. Classic study of Houdini's incredible magic, exposing closely-kept professional secrets and revealing, in general terms, the whole art of stage magic. 67 illustrations. 279pp. 5⅜ x 8½.   22913-0 Pa. $4.00

HOFFMANN'S MODERN MAGIC, Professor Hoffmann. One of the best, and best-known, magicians' manuals of the past century. Hundreds of tricks from card tricks and simple sleight of hand to elaborate illusions involving construction of complicated machinery. 332 illustrations. 563pp. 5⅜ x 8½.   23623-4 Pa. $6.95

THOMAS NAST'S CHRISTMAS DRAWINGS, Thomas Nast. Almost all Christmas drawings by creator of image of Santa Claus as we know it, and one of America's foremost illustrators and political cartoonists. 66 illustrations. 3 illustrations in color on covers. 96pp. 8⅜ x 11¼.
23660-9 Pa. $3.50

FRENCH COUNTRY COOKING FOR AMERICANS, Louis Diat. 500 easy-to-make, authentic provincial recipes compiled by former head chef at New York's Fitz-Carlton Hotel: onion soup, lamb stew, potato pie, more. 309pp. 5⅜ x 8½.   23665-X Pa. $3.95

SAUCES, FRENCH AND FAMOUS, Louis Diat. Complete book gives over 200 specific recipes: bechamel, Bordelaise, hollandaise, Cumberland, apri-cot, etc. Author was one of this century's finest chefs, originator of vichyssoise and many other dishes. Index. 156pp. 5⅜ x 8.
23663-3 Pa. $2.75

TOLL HOUSE TRIED AND TRUE RECIPES, Ruth Graves Wakefield. Authentic recipes from the famous Mass. restaurant: popovers, veal and ham loaf, Toll House baked beans, chocolate cake crumb pudding, much more. Many helpful hints. Nearly 700 recipes. Index. 376pp. 5⅜ x 8½.
23560-2 Pa. $4.95

ILLUSTRATED GUIDE TO SHAKER FURNITURE, Robert Meader. Director, Shaker Museum, Old Chatham, presents up-to-date coverage of all furniture and appurtenances, with much on local styles not available elsewhere. 235 photos. 146pp. 9 x 12. 22819-3 Pa. $6.95

COOKING WITH BEER, Carole Fahy. Beer has as superb an effect on food as wine, and at fraction of cost. Over 250 recipes for appetizers, soups, main dishes, desserts, breads, etc. Index. 144pp. 5⅜ x 8½. (Available in U.S. only) 23661-7 Pa. $3.00

STEWS AND RAGOUTS, Kay Shaw Nelson. This international cookbook offers wide range of 108 recipes perfect for everyday, special occasions, meals-in-themselves, main dishes. Economical, nutritious, easy-to-prepare: goulash, Irish stew, boeuf bourguignon, etc. Index. 134pp. 5⅜ x 8½. 23662-5 Pa. $3.95

DELICIOUS MAIN COURSE DISHES, Marian Tracy. Main courses are the most important part of any meal. These 200 nutritious, economical recipes from around the world make every meal a delight. "I . . . have found it so useful in my own household,"—*N.Y. Times.* Index. 219pp. 5⅜ x 8½. 23664-1 Pa. $3.95

FIVE ACRES AND INDEPENDENCE, Maurice G. Kains. Great back-to-the-land classic explains basics of self-sufficient farming: economics, plants, crops, animals, orchards, soils, land selection, host of other necessary things. Do not confuse with skimpy faddist literature; Kains was one of America's greatest agriculturalists. 95 illustrations. 397pp. 5⅜ x 8½. 20974-1 Pa. **$4.95**

A PRACTICAL GUIDE FOR THE BEGINNING FARMER, Herbert Jacobs. Basic, extremely useful first book for anyone thinking about moving to the country and starting a farm. Simpler than Kains, with greater emphasis on country living in general. 246pp. 5⅜ x 8½. 23675-7 Pa. $3.95

PAPERMAKING, Dard Hunter. Definitive book on the subject by the foremost authority in the field. Chapters dealing with every aspect of history of craft in every part of the world. Over 320 illustrations. 2nd, revised and enlarged (1947) edition. 672pp. 5⅜ x 8½. 23619-6 Pa. $8.95

THE ART DECO STYLE, edited by Theodore Menten. Furniture, jewelry, metalwork, ceramics, fabrics, lighting fixtures, interior decors, exteriors, graphics from pure French sources. Best sampling around. Over 400 photographs. 183pp. 8⅜ x 11¼. 22824-X Pa. $6.95

ACKERMANN'S COSTUME PLATES, Rudolph Ackermann. Selection of 96 plates from the *Repository of Arts,* best published source of costume for English fashion during the early 19th century. 12 plates also in color. Captions, glossary and introduction by editor Stella Blum. Total of 120pp. 8⅜ x 11¼. 23690-0 Pa. $5.00

THE ANATOMY OF THE HORSE, George Stubbs. Often considered the great masterpiece of animal anatomy. Full reproduction of 1766 edition, plus prospectus; original text and modernized text. 36 plates. Introduction by Eleanor Garvey. 121pp. 11 x 14¾.   23402-9 Pa. **$8.95**

BRIDGMAN'S LIFE DRAWING, George B. Bridgman. More than 500 illustrative drawings and text teach you to abstract the body into its major masses, use light and shade, proportion; as well as specific areas of anatomy, of which Bridgman is master. 192pp. 6½ x 9¼. (Available in U.S. only)
        22710-3 Pa. **$4.50**

ART NOUVEAU DESIGNS IN COLOR, Alphonse Mucha, Maurice Verneuil, Georges Auriol. Full-color reproduction of *Combinaisons ornementales* (c. 1900) by Art Nouveau masters. Floral, animal, geometric, interlacings, swashes—borders, frames, spots—all incredibly beautiful. 60 plates, hundreds of designs. 9⅜ x 8-1/16.   22885-1 Pa. **$4.50**

FULL-COLOR FLORAL DESIGNS IN THE ART NOUVEAU STYLE, E. A. Seguy. 166 motifs, on 40 plates, from *Les fleurs et leurs applications decoratives* (1902): borders, circular designs, repeats, allovers, "spots." All in authentic Art Nouveau colors. 48pp. 9⅜ x 12¼.
        23439-8 Pa. **$6.00**

A DIDEROT PICTORIAL ENCYCLOPEDIA OF TRADES AND IN-DUSTRY, edited by Charles C. Gillispie. 485 most interesting plates from the great French Encyclopedia of the 18th century show hundreds of working figures, artifacts, process, land and cityscapes; glassmaking, papermaking, metal extraction, construction, weaving, making furniture, clothing, wigs, dozens, of other activities. Plates fully explained. 920pp. 9 x 12.
        22284-5, 22285-3 Clothbd., Two-vol. set **$50.00**

HANDBOOK OF EARLY ADVERTISING ART, Clarence P. Hornung. Largest collection of copyright-free early and antique advertising art ever compiled. Over 6,000 illustrations, from Franklin's time to the 1890's for special effects, novelty. Valuable source, almost inexhaustible.
Pictorial Volume. Agriculture, the zodiac, animals, autos, birds, Christmas, fire engines, flowers, trees, musical instruments, ships, games and sports, much more. Arranged by subject matter and use. 237 plates. 288pp. 9 x 12.
        20122-8 Clothbd. **$15.00**

Typographical Volume. Roman and Gothic faces ranging from 10 point to 300 point, "Barnum," German and Old English faces, script, logotypes, scrolls and flourishes, 1115 ornamental initials, 67 complete alphabets, more. 310 plates. 320pp. 9 x 12.   20123-6 Clothbd. $15.00

CALLIGRAPHY (CALLIGRAPHIA LATINA), J. G. Schwandner. High point of 18th-century ornamental calligraphy. Very ornate initials, scrolls, borders, cherubs, birds, lettered examples. 172pp. 9 x 13.
        20475-8 Pa. **$7.95**

GEOMETRY, RELATIVITY AND THE FOURTH DIMENSION, Rudolf Rucker. Exposition of fourth dimension, means of visualization, concepts of relativity as Flatland characters continue adventures. Popular, easily followed yet accurate, profound. 141 illustrations. 133pp. 5⅜ x 8½.
23400-2 Pa. $2.75

THE ORIGIN OF LIFE, A. I. Oparin. Modern classic in biochemistry, the first rigorous examination of possible evolution of life from nitrocarbon compounds. Non-technical, easily followed. Total of 295pp. 5⅜ x 8½.
60213-3 Pa. $5.95

PLANETS, STARS AND GALAXIES, A. E. Fanning. Comprehensive introductory survey: the sun, solar system, stars, galaxies, universe, cosmology; quasars, radio stars, etc. 24pp. of photographs. 189pp. 5⅜ x 8½. (Available in U.S. only)
21680-2 Pa. $3.75

THE THIRTEEN BOOKS OF EUCLID'S ELEMENTS, translated with introduction and commentary by Sir Thomas L. Heath. Definitive edition. Textual and linguistic notes, mathematical analysis, 2500 years of critical commentary. Do not confuse with abridged school editions. Total of 1414pp. 5⅜ x 8½.
60088-2, 60089-0, 60090-4 Pa., Three-vol. set $19.50

Prices subject to change without notice.

Available at your book dealer or write for free catalogue to Dept. GI, Dover Publications, Inc., 31 East 2nd St. Mineola., N.Y. 11501. Dover publishes more than 175 books each year on science, elementary and advanced mathematics, biology, music, art, literary history, social sciences and other areas.